The Skinny on Least Squares and Polynomial Curve Fitting

Richard A. Eckhart

Dedication

This book is dedicated to the memory of Charles H. Hoppes, Lehighton (PA) High School math teacher and friend, born November 17, 1909, died October 30, 1988.

Acknowledgement

Thanks to my wife, Mary Jane, who, for a number of years, has encouraged me to 'write a book'. Thanks also to family members and friends for their encouragement in undertaking this task.
"The heavens declare the glory of God; the skies proclaim the work of His hands." Psalm 19:1
Don't Let Me Miss the Glory!

All inquiries should be addressed to:
Richard A. Eckhart, Ph.D.
1703 Claire Street
Prescott, AZ 86301
Email: reckhart@sonshinecs.com

ISBN-13: 978-1483919300

ISBN-10: 1483919307

CONTENTS

CHAPTER 1: INTRODUCTION

THE SKINNY

On TheFreeDictionary.com web site (6), 'skinny' is defined as an adjective meaning very thin, or of, relating to, or resembling skin, and as a slang term meaning inside information, or the real facts. This book gives you the skinny…the inside information, the real facts…about the method of least squares and polynomial curve fitting in a skinny…very thin, compact…format.

This is a 'How to…' book presented in a format resembling a cookbook: do this, this, and this, and you will have the desired result. As a consequence, mathematical derivations, proofs, theory, etc. are relegated to the Appendix.

GENERAL BACKGROUND AND REVIEW

Curve fitting is generally of interest for one of two reasons:

(1) to allow the interpolation of an array or table of data; or

(2) to allow the extrapolation of those data.

Interpolation is the process of finding a value within the domain of the given data while extrapolation is the process of finding a value beyond or outside of the domain of the given data.

An example of interpolation might be the determination of the boiling points of various fractions in some petroleum product from boiling point data obtained from laboratory analysis. Laboratory data might give the boiling points of the petroleum product at 10, 20, 30, 50, 60, 80, and 95 percent distilled, for example, and it is desired to find (estimate) the boiling points at 14, 25, 45, 75, and 90 percent distilled. Note that the desired percents distilled are all within the range of the laboratory data. See Chapter 7: Applications for an example of this type of calculation.

An example of extrapolation of data might be the forecasting of the number of checks a bank will have to process in the next six months based on check processing information from the past two years. Note that the desired information is outside of or beyond the range of the data on which the forecast is based. See Chapter 7: Applications for an example of this type of calculation.

In either case, we are faced with making the decision of whether the data need to be represented with a function that must reproduce each of the original data points exactly, or if the data may be represented by a smoothed curve where each data point is not necessarily represented exactly, but where the general trend of the data is of primary significance.

There are several techniques that might be used for either interpolation or extrapolation of the data, including piece-wise methods, where only a select number of data points are used to make the interpolation/extrapolation, and fitting the data with analytical functions such as logarithmic or exponential mathematical relationships.

In this book we focus on fitting data with an analytical polynomial in two dimensions, such that the original data and its curve fit can be represented graphically in a rectangular coordinate system. More specifically, we focus on the method of least squares where it is desired to represent the entire data array with a single smooth analytical function.

LEAST SQUARES METHOD

Development of the least squares method is generally credited to Carl Friedrich Gauss in the late 18th century (7), so the method is certainly not new. The method finds the best line to represent a set of data points such that the sum of the squares of the vertical distances of each data point from the proposed line is a minimum.

Discussion of the method is frequently limited to a discussion of the use of least squares for establishing a linear trend or straight-line model of the data. In this book, we show how the least squares criterion can be used to find the best quadratic and cubic polynomial data models in addition to the linear model. Higher-order polynomial models also follow from the development presented here, although higher than third-order (cubic) may be unjustified unless one wants an exact representation of the raw data, as we will show later.

The method is frequently described or classified as a regression analysis technique. Regression implies the application of statistical methods in its analysis. However, as we show in the development of the mathematical relationships we use in this book, there are no statistical methods applied to the development. Indeed, the mathematical relationships derive from a straight-forward application of differential calculus methods and techniques with no application of any statistical methods.

We limit our discussion in this book to systems of two-dimensional data; that is, to systems of data that can be represented on a graph with rectangular coordinates with one independent and one dependent variable. In general, in such systems, the independent variable is represented by the letter x, normally plotted along the horizontal axis of the coordinate system, while the dependent variable is represented by the letter y, normally plotted along the vertical axis of the coordinate system.

We also note that our discussion in this book deals with the fitting of data with analytical/polynomial functions of the form

$y = a_0 + a_1 x + a_2 x^2 + ... + a_n x^n$, where x represents the independent variable, y represents the dependent variable, the a_n's are constants, and the exponents of x are positive integers.

IMPLEMENTATION WITH MICROSOFT® EXCEL

There are a number of ways of implementing the methods and calculations we discuss in this book. However, we have chosen to demonstrate the implementation of the calculations with Microsoft® Excel since Excel is readily available on most PC's as one of the components of the Microsoft® Office Suite of applications.

We have included screen shots of Excel spreadsheets to show enough detail so that the reader can easily reproduce the calculations. The graphs we have included have also been generated using Excel. An extensive familiarity with Excel is not required to use this book and to understand the screen shots. However, familiarity with some fundamental concepts of Excel is assumed and helpful, such as the following:

cell addresses: cells in Excel are referenced by the intersection of the column and row that contains the cell in the spreadsheet. For example, the cell address C10 refers to the cell located at the intersection of column C and row 10.

range of cells: cells in Excel can be referenced as a group called a range of cells. A group of cells that is all in the same column will have the same column identifier (ID) (for example, B5:B25) while a group of cells that is all in the same row will have the same row ID (for example, C5:J5). A range of cells having different beginning and ending cell addresses defines a rectangular block of cells (for example, K3:N10).

mathematical operators: the symbol + indicates the mathematical operation of addition, the symbol – indicates the mathematical operation of subtraction, the symbol * indicates the operation of multiplication, the symbol / indicates the operation of division, and the symbol ^ indicates the operation of exponentiation.

relative cell addresses: non-restricted cell addresses are relative, meaning that if a formula in one cell is copied-and-pasted to another cell, the cell address will change relative to the source and destination cell locations. For example, the cell address C10 is a relative cell address.

absolute cell addresses: cell addresses that are modified by the use of the $ symbol are absolute and do not change if copied-and-pasted from one cell to another cell relative to the source and destination cell locations. Either the column or the row or both may be modified with the $ symbol. For example, the cell address D7 is a relative address while the cell address $D$7 is an absolute cell address.

For additional details about each of these fundamental Excel concepts, refer to the Help system in Excel.

WHERE TO GO FROM HERE

So, where do you go from here? Well, that depends on what kind of data you have and what kind of curve you want to use to fit those data. The following tells you where to go next in this book for what you want to do:

If you have three or more data points that you want to fit with a linear (straight-line) equation, go to:

Chapter 2: How To Fit Three or More Points with a Linear Polynomial

Use this procedure to represent data that is constantly increasing/rising or constantly decreasing/falling in the dependent variable (y). The mathematical representation of the data will be in the form of the equation $y = a_0 + a_1 x$.

If you have four or more data points that you want to fit with a quadratic equation, go to:

Chapter 3: How To Fit Four or More Points with a Quadratic Polynomial

Use this procedure to represent data that could exhibit a single maximum or a single minimum value in the dependent variable (y). The mathematical representation of the data will be in the form of the equation $y = a_0 + a_1 x + a_2 x^2$.

If you have five or more data points that you want to fit with a cubic equation, go to:

Chapter 4: How To Fit Five or More Points with a Cubic Polynomial

Use this procedure to represent data that could exhibit both a maximum and a minimum value in the dependent variable (y). The mathematical representation of the data will be in the form of the equation

$y = a_0 + a_1 x + a_2 x^2 + a_3 x^3$.

If you have any number of data points that you want to fit with a polynomial equation that goes exactly through each and every data point, go to:

Chapter 5: How To Fit n Points with an (n-1)th Order Polynomial

Use this procedure to represent the exact values of the given data. The mathematical representation of the data will be in the form of the equation

$y = a_0 + a_1 x + a_2 x^2 + \ldots + a_n x^n$ where the number of data points is $(n+1)$.

Implementation of each procedure is illustrated with screen shots taken from Microsoft® Excel.

CHAPTER 2: HOW TO FIT THREE OR MORE POINTS WITH A LINEAR POLYNOMIAL

WEB VIDEO ADDRESS:
WWW.SONSHINECS.COM/VIDEOS/LINEAR_LS_VIDEO.MP4

Here is a step-by-step outline for finding the best straight-line, linear, first-order polynomial fit of a set of two-dimensional data points based on the criterion of the method of least squares. The procedure described is implemented in a Microsoft® Excel spreadsheet, using appropriate Excel mathematical functions, and is demonstrated using a set of five data points. The procedure can be extended and applied to any number of data points (three or more). In the procedure described, the variable x represents the independent variable, normally plotted along the horizontal axis in a rectangular coordinate system, and y represents the dependent variable, normally plotted along the vertical axis in a rectangular coordinate system. The resulting linear equation is of the form $y = a_0 + a_1 x$, where a_0 and a_1 are constants.

Step 1. Enter the given x, y data into adjacent cells in Excel.

x	y
0.0	1.0
1.0	3.0
2.0	2.0
3.0	4.0
4.0	5.0

Figure 2.1

Step 2. Count the number of x, y-pairs using the Excel ‘Count()’ function.

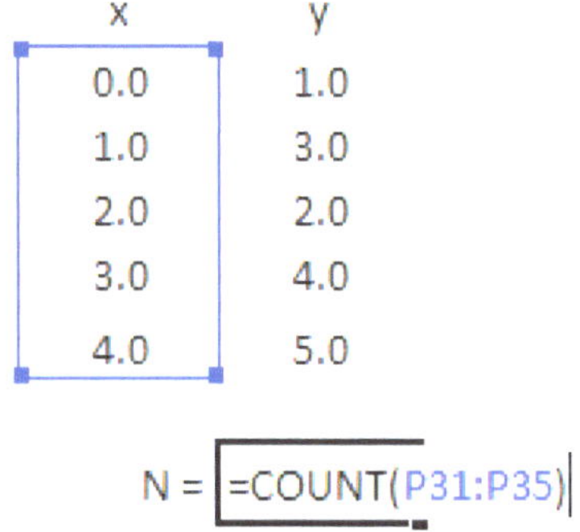

x	y
0.0	1.0
1.0	3.0
2.0	2.0
3.0	4.0
4.0	5.0

N = =COUNT(P31:P35)

Figure 2.2

Step 3. Compute the sum of the x-values and the sum of the y-values using the Excel ‘Sum()’ function.

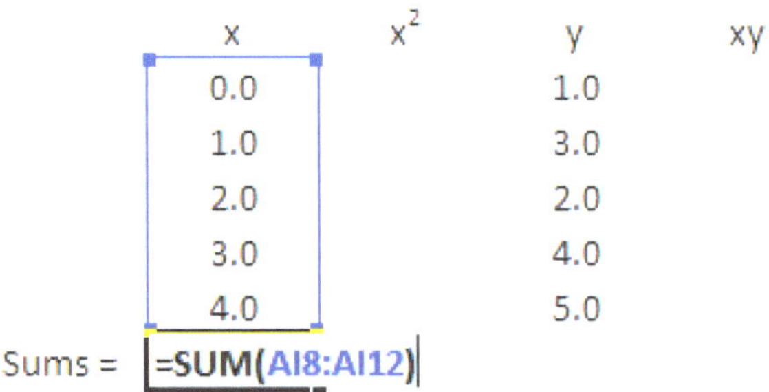

	x	x^2	y	xy
	0.0		1.0	
	1.0		3.0	
	2.0		2.0	
	3.0		4.0	
	4.0		5.0	
Sums =	=SUM(AI8:AI12)			

Figure 2.3

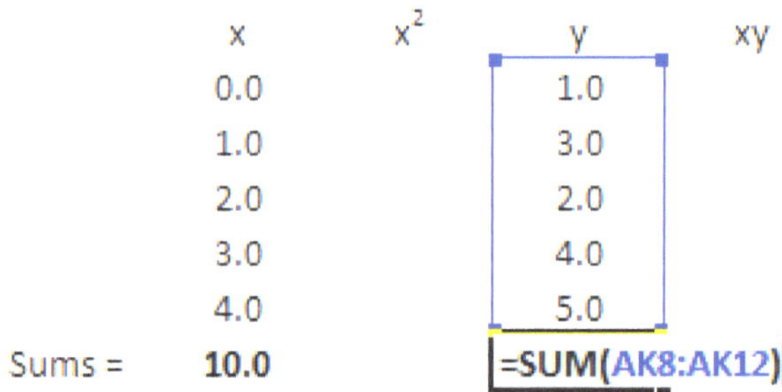

	x	x^2	y	xy
	0.0		1.0	
	1.0		3.0	
	2.0		2.0	
	3.0		4.0	
	4.0		5.0	
Sums =	**10.0**		**=SUM(AK8:AK12)**	

Figure 2.4

	x	x^2	y	xy
	0.0		1.0	
	1.0		3.0	
	2.0		2.0	
	3.0		4.0	
	4.0		5.0	
Sums =	**10.0**		**15.0**	

Figure 2.5

Step 4. Square each x-value (multiply each x-value by itself) and compute the sum of the squares using the Excel 'Sum()' function.

	x	x^2	y	xy
	0.0	=AI8*AI8	1.0	
	1.0		3.0	
	2.0		2.0	
	3.0		4.0	
	4.0		5.0	
Sums =	**10.0**		**15.0**	

Figure 2.6

	x	x^2	y	xy
	0.0	0.0	1.0	
	1.0	1.0	3.0	
	2.0	4.0	2.0	
	3.0	9.0	4.0	
	4.0	16.0	5.0	
Sums =	**10.0**	**=SUM(AJ8:AJ12)**		

Figure 2.7

	x	x^2	y	xy
	0.0	0.0	1.0	
	1.0	1.0	3.0	
	2.0	4.0	2.0	
	3.0	9.0	4.0	
	4.0	16.0	5.0	
Sums =	**10.0**	**30.0**	**15.0**	

Figure 2.8

Step 5. Multiply each x-value by its corresponding y-value and compute the sum of the products using the Excel 'Sum()' function.

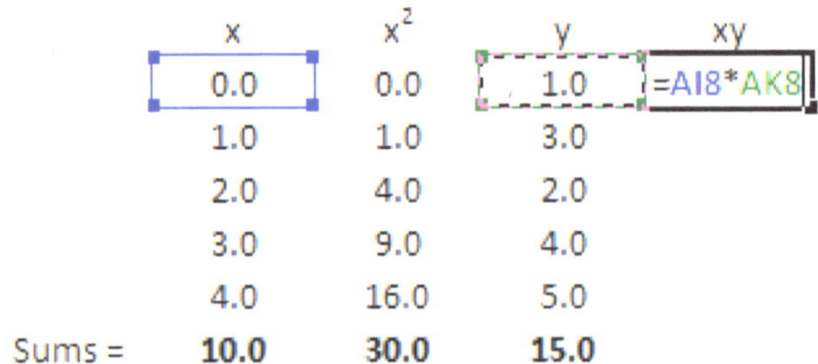

	x	x^2	y	xy
	0.0	0.0	1.0	=AI8*AK8
	1.0	1.0	3.0	
	2.0	4.0	2.0	
	3.0	9.0	4.0	
	4.0	16.0	5.0	
Sums =	**10.0**	**30.0**	**15.0**	

Figure 2.9

	x	x^2	y	xy
	0.0	0.0	1.0	0.0
	1.0	1.0	3.0	3.0
	2.0	4.0	2.0	4.0
	3.0	9.0	4.0	12.0
	4.0	16.0	5.0	20.0
Sums =	**10.0**	**30.0**	**15.0**	**=SUM(AL8:AL12)**

Figure 2.10

	x	x^2	y	xy
	0.0	0.0	1.0	0.0
	1.0	1.0	3.0	3.0
	2.0	4.0	2.0	4.0
	3.0	9.0	4.0	12.0
	4.0	16.0	5.0	20.0
Sums =	**10.0**	**30.0**	**15.0**	**39.0**

Figure 2.11

The results of these calculations are shown in the Excel screen shot below.

x	y		x	x^2	y	xy
0.0	1.0		0.0	0.0	1.0	0.0
1.0	3.0		1.0	1.0	3.0	3.0
2.0	2.0		2.0	4.0	2.0	4.0
3.0	4.0		3.0	9.0	4.0	12.0
4.0	5.0		4.0	16.0	5.0	20.0
		Sums =	**10.0**	**30.0**	**15.0**	**39.0**
N =	**5**					

COUNT() Function

SUM() Function

Figure 2.12

Step 6. Put the Count() value and the SUM() values into a 2-by-2 (2 rows by 2 columns) and a 2-by-1 (2 rows by 1 column) set of Excel cells as shown in the Excel screen shot below to create the **A** matrix and the **B** vector.

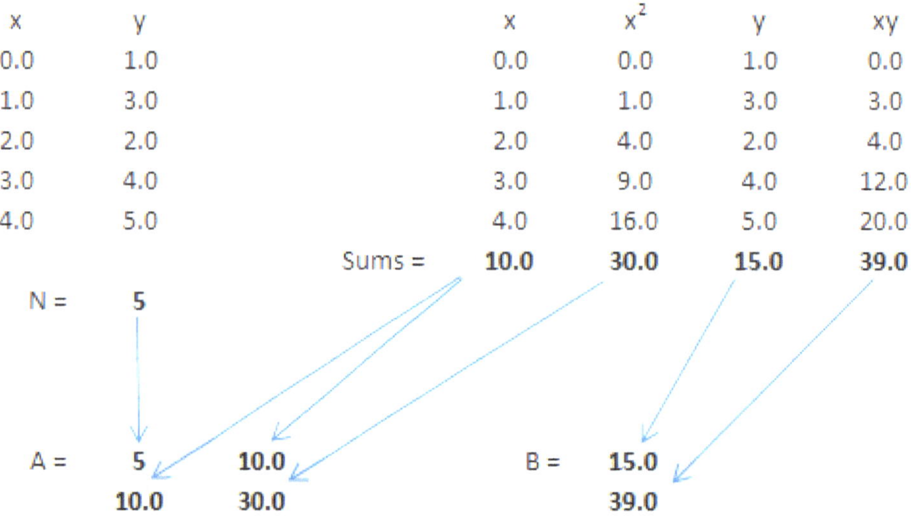

x y
0.0 1.0
1.0 3.0
2.0 2.0
3.0 4.0
4.0 5.0
x x2 y xy
0.0 0.0 1.0 0.0
1.0 1.0 3.0 3.0
2.0 4.0 2.0 4.0
3.0 9.0 4.0 12.0
4.0 16.0 5.0 20.0
Sums = 10.0 30.0 15.0 39.0
N = 5
A = 5 10.0
10.0 30.0
B = 15.0
39.0

Figure 2.13

Step 7. Calculate the inverse of the $\mathbf{A}$ matrix ($\mathbf{A}^{-1}$) by marking off a 2-by-2 array of cells in Excel.

A = 5 10.0
10.0 30.0

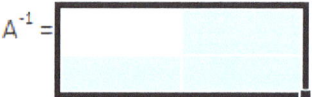

Figure 2.14

Enter into the first cell of the $\mathbf{A}^{-1}$ matrix the Excel 'MINVERSE()' function. To enter the function, type '=MINVERSE(range)', where 'range' is the range of cells of the $\mathbf{A}$ matrix, then hold down the 'Ctrl' and 'Shift' keys and press 'Enter'.

A =	5	10.0
	10.0	30.0

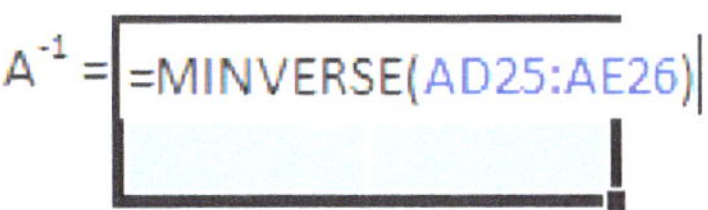

Figure 2.15

Hold down the <Ctrl> and <Shift> keys and press <Enter>.

A =	5	10.0
	10.0	30.0

A^{-1} =	0.60	-0.20
	-0.20	0.10

Figure 2.16

Step 8. Calculate the coefficients of the linear least squares curve fit by marking off a 2-by-1 array of cells in Excel.

A^{-1} =	0.60	-0.20
	-0.20	0.10

X =

B =	15.0
	39.0

Figure 2.17

Enter into the first cell of the **X** vector the Excel 'MMULT()' function. To enter the function, type '=MMULT(range1,range2)', where 'range1' is the range of cells of the $\mathbf{A}^{-1}$ matrix and 'range2' is the range of cells of the **B** vector, then hold down the 'Ctrl' and 'Shift' keys and press 'Enter'.

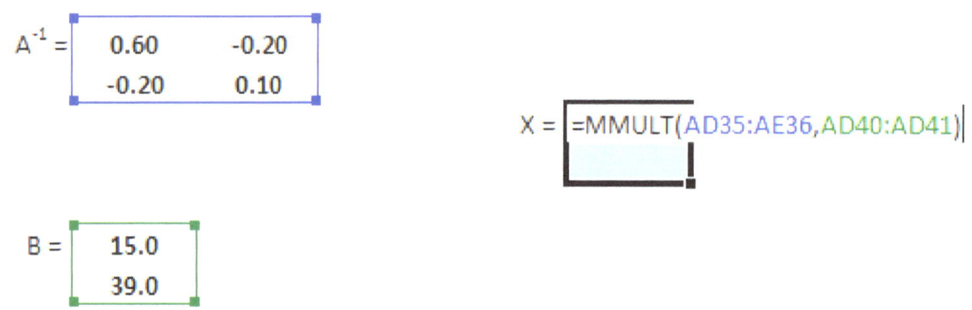

Figure 2.18

Hold down the <Ctrl> and <Shift> keys and press <Enter>.

A^{-1} =	0.60	-0.20
	-0.20	0.10

X =	1.20	= a_0
	0.90	= a_1

B =	15.0
	39.0

Figure 2.19

The **X** vector contains the coefficients of the best linear curve fit of the given data of the form $y = a_0 + a_1 x$, based on the least squares criterion. In this example, the best linear curve fit of the given data, based on the least squares criterion, is $y = 1.2 + 0.9x$. This is the straight-line equation that minimizes the sum of the squares of the differences for the curve fit.

We can confirm this fit by using Excel to draw a graph of the original data and the linear curve fit.

x	y	y_{calc}
0.0	1.0	1.20
1.0	3.0	2.10
2.0	2.0	3.00
3.0	4.0	3.90
4.0	5.0	4.80

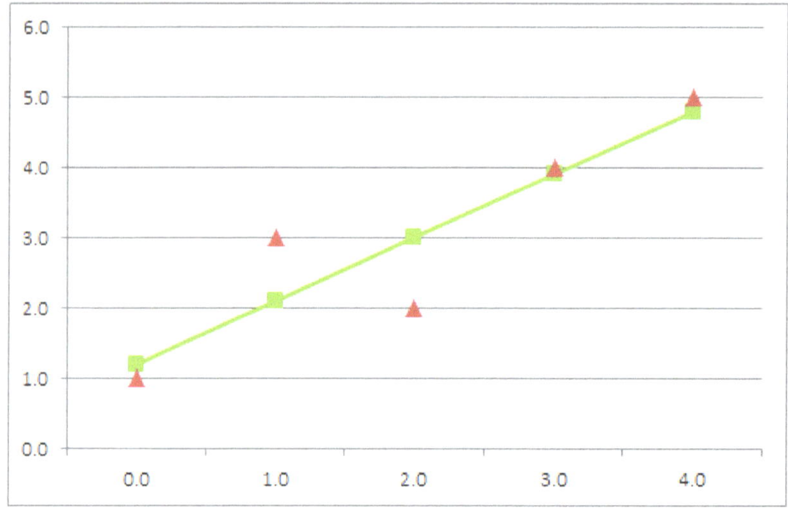

Figure 2.20

Here, the table shows the original x, y data along with the y-values calculated from the linear curve fit equation (y_{calc}). The triangular points on the graph are the original (5) data points while the squares are the calculated y-values corresponding to the given x-values. The y-intercept of the line is 1.2 and the slope of the line is

$$\frac{4.80 - 1.20}{4 - 0} = \frac{3.60}{4} = 0.9.$$

CHAPTER 3: HOW TO FIT FOUR OR MORE POINTS WITH A QUADRATIC POLYNOMIAL

WEB VIDEO ADDRESS:
WWW.SONSHINECS.COM/VIDEOS/QUADRATIC_LS_VIDEO.MP4

Here is a step-by-step outline of how to find the best quadratic, second-order polynomial fit of a set of two-dimensional data points based on the criterion of the method of least squares. As in the procedure for the linear curve fit described above, this procedure is implemented in a Microsoft® Excel spreadsheet, using appropriate Excel mathematical functions, and is demonstrated using the same set of five data points as in the linear procedure. The procedure can be extended and applied to any number of data points (four or more). In the procedure described, the variable x represents the independent variable, normally plotted along the horizontal axis in a rectangular coordinate system, and the variable y represents the dependent variable, normally plotted along the vertical axis in a rectangular coordinate system. The resulting quadratic equation is of the form $y = a_0 + a_1 x + a_2 x^2$, where a_0, a_1, and a_2 are constants.

Step 1. Enter the given x, y data into adjacent cells in Excel.

x	y
0.0	1.0
1.0	3.0
2.0	2.0
3.0	4.0
4.0	5.0

Figure 3.1

Step 2. Count the number of x, y-pairs using the Excel 'Count()' function.

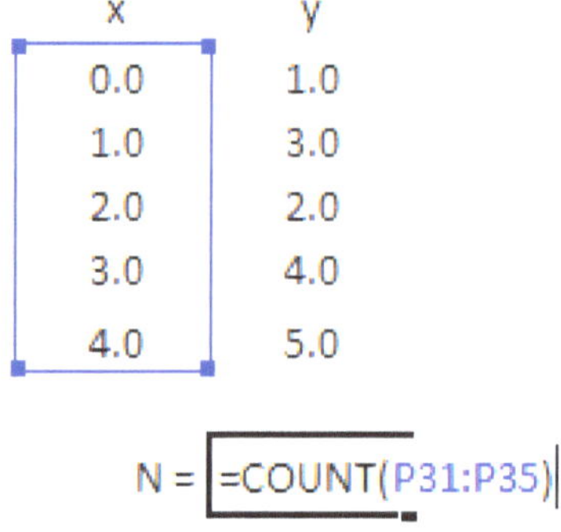

x	y
0.0	1.0
1.0	3.0
2.0	2.0
3.0	4.0
4.0	5.0

N = =COUNT(P31:P35)

Figure 3.2

Step 3. Compute the sum of the x-values and the sum of the y-values using the Excel 'Sum()' function.

	x	x^2	x^3	x^4	y	xy	x^2y
	0.0				1.0		
	1.0				3.0		
	2.0				2.0		
	3.0				4.0		
	4.0				5.0		
Sums =	=SUM(R8:R12)						

Figure 3.3

	x	x^2	x^3	x^4	y	xy	x^2y
	0.0				1.0		
	1.0				3.0		
	2.0				2.0		
	3.0				4.0		
	4.0				5.0		
Sums =	**10.0**				=SUM(V8:V12)		

Figure 3.4

	x	x^2	x^3	x^4	y	xy	x^2y
	0.0				1.0		
	1.0				3.0		
	2.0				2.0		
	3.0				4.0		
	4.0				5.0		
Sums =	**10.0**				**15.0**		

Figure 3.5

Step 4. Square each x-value (multiply each x-value by itself) and compute the sum of the squares using the Excel 'Sum()' function.

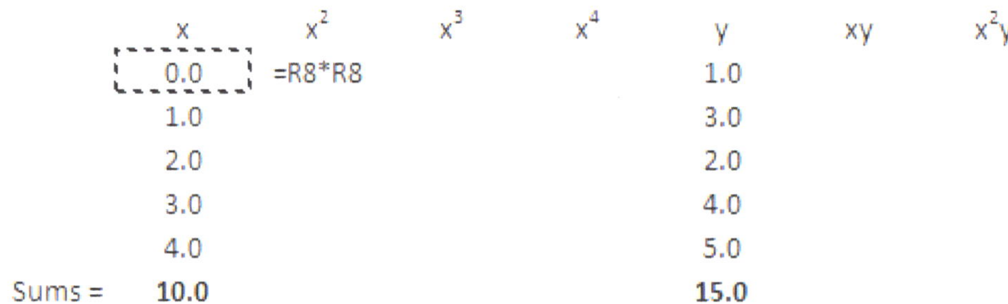

	x	x^2	x^3	x^4	y	xy	x^2y
	0.0	=R8*R8			1.0		
	1.0				3.0		
	2.0				2.0		
	3.0				4.0		
	4.0				5.0		
Sums =	**10.0**				**15.0**		

Figure 3.6

	x	x^2	x^3	x^4	y	xy	x^2y
	0.0	0.0			1.0		
	1.0	1.0			3.0		
	2.0	4.0			2.0		
	3.0	9.0			4.0		
	4.0	16.0			5.0		
Sums =	**10.0**	=SUM(S8:S12)			**15.0**		

Figure 3.7

	x	x^2	x^3	x^4	y	xy	x^2y
	0.0	0.0			1.0		
	1.0	1.0			3.0		
	2.0	4.0			2.0		
	3.0	9.0			4.0		
	4.0	16.0			5.0		
Sums =	**10.0**	**30.0**			**15.0**		

Figure 3.8

Step 5. Calculate each x^3 and each x^4 and compute the sum of the x^3's and the sum of the x^4's using the Excel 'Sum()' function.

	x	x^2	x^3	x^4	y	xy	x^2y
	0.0		=R8*R8*R8		1.0		
	1.0	1.0			3.0		
	2.0	4.0			2.0		
	3.0	9.0			4.0		
	4.0	16.0			5.0		
Sums =	**10.0**	**30.0**			**15.0**		

Figure 3.9

	x	x^2	x^3	x^4	y	xy	x^2y
	0.0	0.0	0.0	=R8*R8*R8*R8			
	1.0	1.0	1.0		3.0		
	2.0	4.0	8.0		2.0		
	3.0	9.0	27.0		4.0		
	4.0	16.0	64.0		5.0		
Sums =	**10.0**	**30.0**			**15.0**		

Figure 3.10

	x	x^2	x^3	x^4	y	xy	x^2y
	0.0	0.0	0.0	0.0	1.0		
	1.0	1.0	1.0	1.0	3.0		
	2.0	4.0	8.0	16.0	2.0		
	3.0	9.0	27.0	81.0	4.0		
	4.0	16.0	64.0	256.0	5.0		
Sums =	**10.0**	**30.0**	=SUM(T8:T12)		**15.0**		

Figure 3.11

	x	x^2	x^3	x^4	y	xy	x^2y
	0.0	0.0	0.0	0.0	1.0		
	1.0	1.0	1.0	1.0	3.0		
	2.0	4.0	8.0	16.0	2.0		
	3.0	9.0	27.0	81.0	4.0		
	4.0	16.0	64.0	256.0	5.0		
Sums =	**10.0**	**30.0**	**100.0**	=SUM(U8:U12)			

Figure 3.12

	x	x^2	x^3	x^4	y	xy	x^2y
	0.0	0.0	0.0	0.0	1.0		
	1.0	1.0	1.0	1.0	3.0		
	2.0	4.0	8.0	16.0	2.0		
	3.0	9.0	27.0	81.0	4.0		
	4.0	16.0	64.0	256.0	5.0		
Sums =	**10.0**	**30.0**	**100.0**	**354.0**	**15.0**		

Figure 3.13

Step 6. Calculate each xy product and each x^2y product and compute the sum of the xy's and the sum of the x^2y's using the Excel 'Sum()' function.

	x	x^2	x^3	x^4	y	xy	x^2y
	0.0	0.0	0.0	0.0	1.0	=R8*V8	
	1.0	1.0	1.0	1.0	3.0		
	2.0	4.0	8.0	16.0	2.0		
	3.0	9.0	27.0	81.0	4.0		
	4.0	16.0	64.0	256.0	5.0		
Sums =	**10.0**	**30.0**	**100.0**	**354.0**	**15.0**		

Figure 3.14

	x	x^2	x^3	x^4	y	xy	x^2y
	0.0	0.0	0.0	0.0	1.0	0.0	=R8*R8*V8
	1.0	1.0	1.0	1.0	3.0	3.0	
	2.0	4.0	8.0	16.0	2.0	4.0	
	3.0	9.0	27.0	81.0	4.0	12.0	
	4.0	16.0	64.0	256.0	5.0	20.0	
Sums =	**10.0**	**30.0**	**100.0**	**354.0**	**15.0**		

Figure 3.15

	x	x^2	x^3	x^4	y	xy	x^2y
	0.0	0.0	0.0	0.0	1.0	0.0	0.0
	1.0	1.0	1.0	1.0	3.0	3.0	3.0
	2.0	4.0	8.0	16.0	2.0	4.0	8.0
	3.0	9.0	27.0	81.0	4.0	12.0	36.0
	4.0	16.0	64.0	256.0	5.0	20.0	80.0
Sums =	**10.0**	**30.0**	**100.0**	**354.0**	**15.0**	=SUM(W8:W12)	

Figure 3.16

	x	x^2	x^3	x^4	y	xy	x^2y
	0.0	0.0	0.0	0.0	1.0	0.0	0.0
	1.0	1.0	1.0	1.0	3.0	3.0	3.0
	2.0	4.0	8.0	16.0	2.0	4.0	8.0
	3.0	9.0	27.0	81.0	4.0	12.0	36.0
	4.0	16.0	64.0	256.0	5.0	20.0	80.0
Sums =	**10.0**	**30.0**	**100.0**	**354.0**	**15.0**	**39.0**	=SUM(X8:X12)

Figure 3.17

	x	x^2	x^3	x^4	y	xy	x^2y
	0.0	0.0	0.0	0.0	1.0	0.0	0.0
	1.0	1.0	1.0	1.0	3.0	3.0	3.0
	2.0	4.0	8.0	16.0	2.0	4.0	8.0
	3.0	9.0	27.0	81.0	4.0	12.0	36.0
	4.0	16.0	64.0	256.0	5.0	20.0	80.0
Sums =	**10.0**	**30.0**	**100.0**	**354.0**	**15.0**	**39.0**	**127.0**

Figure 3.18

Step 7. Put the Count() value and the SUM() values into a 3-by-3 (3 rows by 3 columns) and a 3-by-1 (3 rows by 1 column) set of Excel cells as shown in the Excel screen shot below to create the $\mathbf{A}$ matrix and the $\mathbf{B}$ vector.

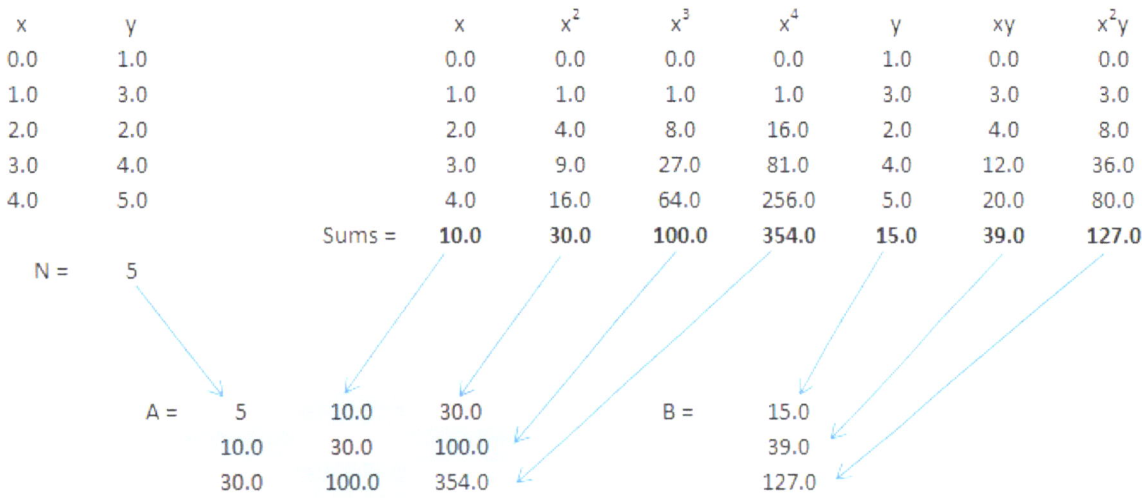

x	y		x	x^2	x^3	x^4	y	xy	x^2y
0.0	1.0		0.0	0.0	0.0	0.0	1.0	0.0	0.0
1.0	3.0		1.0	1.0	1.0	1.0	3.0	3.0	3.0
2.0	2.0		2.0	4.0	8.0	16.0	2.0	4.0	8.0
3.0	4.0		3.0	9.0	27.0	81.0	4.0	12.0	36.0
4.0	5.0		4.0	16.0	64.0	256.0	5.0	20.0	80.0
		Sums =	**10.0**	**30.0**	**100.0**	**354.0**	**15.0**	**39.0**	**127.0**
N =	5								

A =				B =
	5	10.0	30.0	15.0
	10.0	30.0	100.0	39.0
	30.0	100.0	354.0	127.0

Figure 3.19

Step 8. Calculate the inverse of the $\mathbf{A}$ matrix ($\mathbf{A}^{-1}$) by marking off a 3-by-3 array of cells in Excel.

A =			
	5	10.0	30.0
	10.0	30.0	100.0
	30.0	100.0	354.0

A^{-1} =

Figure 3.20

Enter into the first cell of the $\mathbf{A}^{-1}$ matrix the Excel 'MINVERSE()' function. To enter the function, type '=MINVERSE(range)', where 'range' is the range of cells of the $\mathbf{A}$ matrix, then hold down the 'Ctrl' and 'Shift' keys and press 'Enter'.

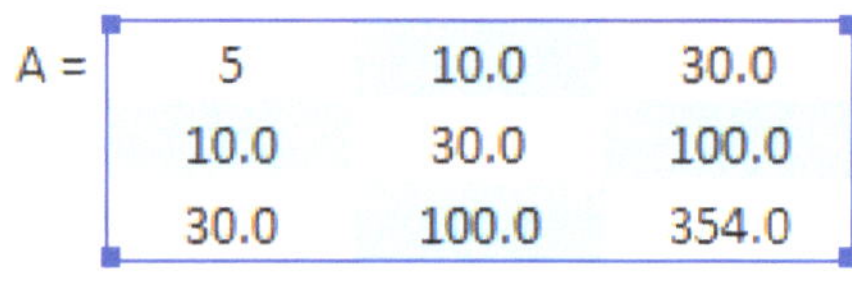

A =			
	5	10.0	30.0
	10.0	30.0	100.0
	30.0	100.0	354.0

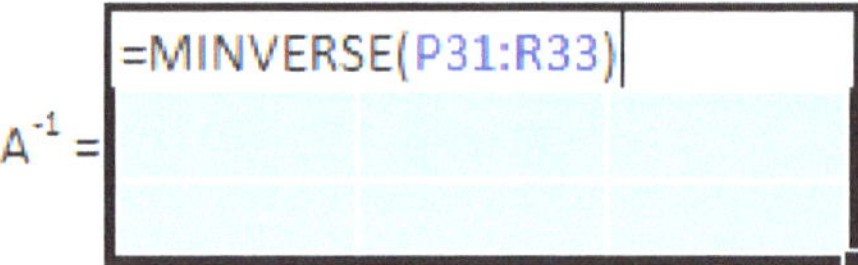

Figure 3.21

Hold down the <Ctrl> and <Shift> keys and press <Enter>.

A =			
	5	10.0	30.0
	10.0	30.0	100.0
	30.0	100.0	354.0

A^{-1} =			
	0.886	-0.771	0.143
	-0.771	1.243	-0.286
	0.143	-0.286	0.071

Figure 3.22

Step 9. Calculate the coefficients of the quadratic least squares curve fit by marking off a 3-by-1 array of cells in Excel.

A =	5	10.0	30.0	B =	15.0
	10.0	30.0	100.0		39.0
	30.0	100.0	354.0		127.0

	0.886	-0.771	0.143		
A^{-1} =	-0.771	1.243	-0.286	X =	
	0.143	-0.286	0.071		

Figure 3.23

Enter into the first cell of the **X** vector the Excel 'MMULT()' function. To enter the function, type '=MMULT(range1,range2)', where 'range1' is the range of cells of the $\mathbf{A}^{-1}$ matrix and 'range2' is the range of cells of the **B** vector, then hold down the 'Ctrl' and 'Shift' keys and press 'Enter'.

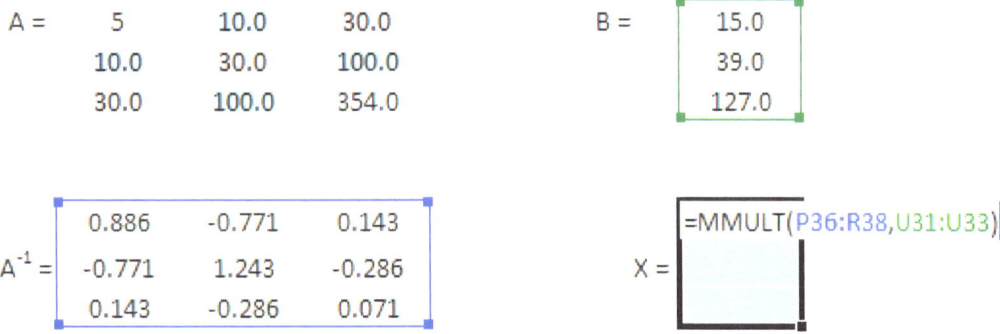

Figure 3.24

Hold down the <Ctrl> and <Shift> keys and press <Enter>.

A =	5	10.0	30.0		B =	15.0	
	10.0	30.0	100.0			39.0	
	30.0	100.0	354.0			127.0	
	0.886	-0.771	0.143			1.343	$= a_0$
A^{-1} =	-0.771	1.243	-0.286		X =	0.614	$= a_1$
	0.143	-0.286	0.071			0.071	$= a_2$

Figure 3.25

The **X** vector contains the coefficients of the best quadratic curve fit of the given data of the form $y = a_0 + a_1 x + a_2 x^2$, based on the least squares criterion. In this example, the best quadratic curve fit of the given data, based on the least squares criterion, is $y = 1.343 + 0.614x + 0.071x^2$. This is the quadratic equation that minimizes the sum of the squares of the differences for the curve fit.

We can confirm this fit by using Excel to draw a graph of the original data and the quadratic curve fit.

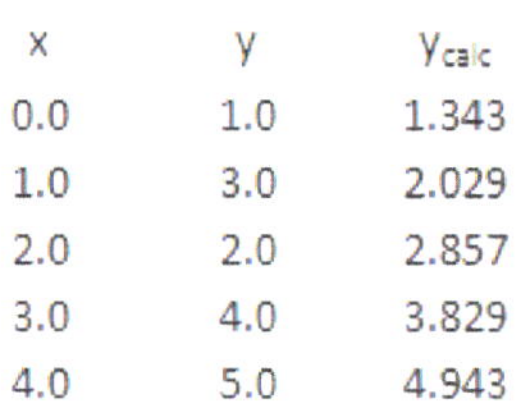

x	y	y_{calc}
0.0	1.0	1.343
1.0	3.0	2.029
2.0	2.0	2.857
3.0	4.0	3.829
4.0	5.0	4.943

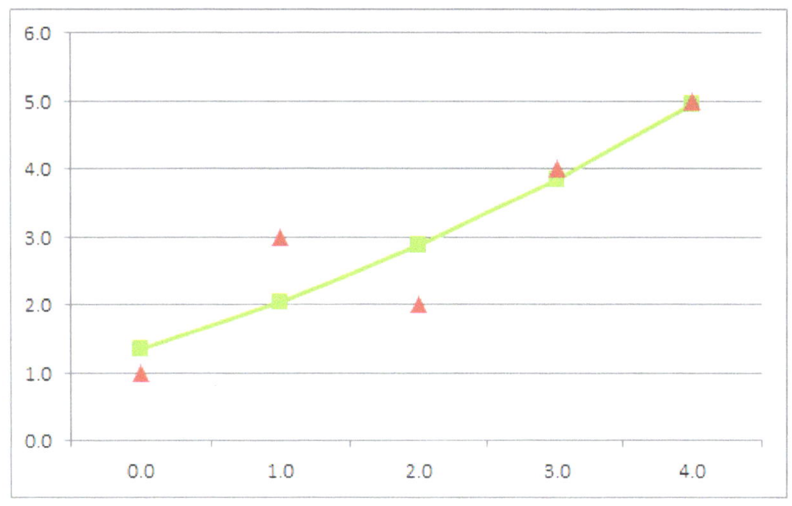

Figure 3.26

Here, the table shows the original x, y data along with the y-values calculated from the quadratic curve fit equation (y_{calc}). The triangular points on the graph are the original (5) data points while the squares are the calculated y-values corresponding to the given x-values. Note that the y-intercept of the line is 1.343 which corresponds to the value of a_0 and the value of y when $x = 0$.

CHAPTER 4: HOW TO FIT FIVE OR MORE POINTS WITH A CUBIC POLYNOMIAL

WEB VIDEO ADDRESS:
WWW.SONSHINECS.COM/VIDEOS/CUBIC_LS_VIDEO.MP4

Here is a step-by-step outline of how to find the best cubic, third-order polynomial fit of a set of two-dimensional data points based on the criterion of the method of least squares. As in the procedures for the linear and quadratic curve fits described above, this procedure is implemented in a Microsoft® Excel spreadsheet using appropriate Excel mathematical functions, and is demonstrated using the same set of five data points as in the linear and quadratic procedures. The procedure can be extended and applied to any number of data points (five or more). In the procedure described, the variable x represents the independent variable, normally plotted along the horizontal axis in a rectangular coordinate system, and y represents the dependent variable, normally plotted along the vertical axis in a rectangular coordinate system. The resulting cubic equation is of the form $y = a_0 + a_1 x + a_2 x^2 + a_3 x^3$, where a_0, a_1, a_2, and a_3 are constants.

Step 1. Enter the given x, y data into adjacent cells in Excel.

x	y
0.0	1.0
1.0	3.0
2.0	2.0
3.0	4.0
4.0	5.0

Figure 4.1

Step 2. Count the number of x, y-pairs using the Excel 'Count()' function.

x	y
0.0	1.0
1.0	3.0
2.0	2.0
3.0	4.0
4.0	5.0

N = =COUNT(P31:P35)

Figure 4.2

Step 3. Compute the sum of the x-values and the sum of the y-values using the Excel 'Sum()' function.

	x	x^2	x^3	x^4	x^5	x^6	y	xy	x^2y	x^3y
	0.0						1.0			
	1.0						3.0			
	2.0						2.0			
	3.0						4.0			
	4.0						5.0			
Sums =	=SUM(R8:R12)									

Figure 4.3

	x	x^2	x^3	x^4	x^5	x^6	y	xy	x^2y	x^3y
	0.0						1.0			
	1.0						3.0			
	2.0						2.0			
	3.0						4.0			
	4.0						5.0			
Sums =	**10.0**						=SUM(X8:X12)			

Figure 4.4

	x	x^2	x^3	x^4	x^5	x^6	y	xy	x^2y	x^3y
	0.0						1.0			
	1.0						3.0			
	2.0						2.0			
	3.0						4.0			
	4.0						5.0			
Sums =	**10.0**						**15.0**			

Figure 4.5

Step 4. Calculate each x^2, x^3, x^4, x^5, and x^6 and the sums of the x^2's, x^3's, x^4's, x^5's, and x^6's using the Excel 'Sum()' function.

	x	x^2	x^3	x^4	x^5	x^6	y	xy	x^2y	x^3y
	0.0	=R8*R8					1.0			
	1.0						3.0			
	2.0						2.0			
	3.0						4.0			
	4.0						5.0			
Sums =	**10.0**						15.0			

Figure 4.6

	x	x^2	x^3	x^4	x^5	x^6	y	xy	x^2y	x^3y
	0.0		=R8*R8*R8				1.0			
	1.0	1.0					3.0			
	2.0	4.0					2.0			
	3.0	9.0					4.0			
	4.0	16.0					5.0			
Sums =	**10.0**						15.0			

Figure 4.7

	x	x^2	x^3	x^4	x^5	x^6	y	xy	x^2y	x^3y
	0.0	0.0	0.0	=R8*R8*R8*R8			1.0			
	1.0	1.0	1.0				3.0			
	2.0	4.0	8.0				2.0			
	3.0	9.0	27.0				4.0			
	4.0	16.0	64.0				5.0			
Sums =	**10.0**						15.0			

Figure 4.8

	x	x^2	x^3	x^4	x^5	x^6	y	xy	x^2y	x^3y
	0.0	0.0	0.0	0.0	=R8*R8*R8*R8*R8		1.0			
	1.0	1.0	1.0	1.0			3.0			
	2.0	4.0	8.0	16.0			2.0			
	3.0	9.0	27.0	81.0			4.0			
	4.0	16.0	64.0	256.0			5.0			
Sums =	**10.0**						15.0			

Figure 4.9

	x	x^2	x^3	x^4	x^5	x^6	y	xy	x^2y	x^3y
	0.0	0.0	0.0	0.0	0.0	=R8*R8*R8*R8*R8*R8				
	1.0	1.0	1.0	1.0	1.0		3.0			
	2.0	4.0	8.0	16.0	32.0		2.0			
	3.0	9.0	27.0	81.0	243.0		4.0			
	4.0	16.0	64.0	256.0	1024.0		5.0			
Sums =	**10.0**						15.0			

Figure 4.10

	x	x^2	x^3	x^4	x^5	x^6	y	xy	x^2y	x^3y
	0.0	0.0	0.0	0.0	0.0	0.0	1.0			
	1.0	1.0	1.0	1.0	1.0	1.0	3.0			
	2.0	4.0	8.0	16.0	32.0	64.0	2.0			
	3.0	9.0	27.0	81.0	243.0	729.0	4.0			
	4.0	16.0	64.0	256.0	1024.0	4096.0	5.0			
Sums =	**10.0**	=SUM(S8:S12)					15.0			

Figure 4.11

	x	x^2	x^3	x^4	x^5	x^6	y	xy	x^2y	x^3y
	0.0	0.0	0.0	0.0	0.0	0.0	1.0			
	1.0	1.0	1.0	1.0	1.0	1.0	3.0			
	2.0	4.0	8.0	16.0	32.0	64.0	2.0			
	3.0	9.0	27.0	81.0	243.0	729.0	4.0			
	4.0	16.0	64.0	256.0	1024.0	4096.0	5.0			
Sums =	**10.0**	**30.0**	=SUM(T8:T12)				15.0			

Figure 4.12

	x	x^2	x^3	x^4	x^5	x^6	y	xy	x^2y	x^3y
	0.0	0.0	0.0	0.0	0.0	0.0	1.0			
	1.0	1.0	1.0	1.0	1.0	1.0	3.0			
	2.0	4.0	8.0	16.0	32.0	64.0	2.0			
	3.0	9.0	27.0	81.0	243.0	729.0	4.0			
	4.0	16.0	64.0	256.0	1024.0	4096.0	5.0			
Sums =	**10.0**	**30.0**	**100.0**	=SUM(U8:U12)			15.0			

Figure 4.13

	x	x^2	x^3	x^4	x^5	x^6	y	xy	x^2y	x^3y
	0.0	0.0	0.0	0.0	0.0	0.0	1.0			
	1.0	1.0	1.0	1.0	1.0	1.0	3.0			
	2.0	4.0	8.0	16.0	32.0	64.0	2.0			
	3.0	9.0	27.0	81.0	243.0	729.0	4.0			
	4.0	16.0	64.0	256.0	1024.0	4096.0	5.0			
Sums =	**10.0**	**30.0**	**100.0**	**354.0**	=SUM(V8:V12)		15.0			

Figure 4.14

	x	x^2	x^3	x^4	x^5	x^6	y	xy	x^2y	x^3y
	0.0	0.0	0.0	0.0	0.0	0.0	1.0			
	1.0	1.0	1.0	1.0	1.0	1.0	3.0			
	2.0	4.0	8.0	16.0	32.0	64.0	2.0			
	3.0	9.0	27.0	81.0	243.0	729.0	4.0			
	4.0	16.0	64.0	256.0	1024.0	4096.0	5.0			
Sums =	**10.0**	**30.0**	**100.0**	**354.0**	**1300.0**	**=SUM(W8:W12)**				

Figure 4.15

	x	x^2	x^3	x^4	x^5	x^6	y	xy	x^2y	x^3y
	0.0	0.0	0.0	0.0	0.0	0.0	1.0			
	1.0	1.0	1.0	1.0	1.0	1.0	3.0			
	2.0	4.0	8.0	16.0	32.0	64.0	2.0			
	3.0	9.0	27.0	81.0	243.0	729.0	4.0			
	4.0	16.0	64.0	256.0	1024.0	4096.0	5.0			
Sums =	**10.0**	**30.0**	**100.0**	**354.0**	**1300.0**	**4890.0**	15.0			

Figure 4.16

Step 5. Calculate each xy, x^2y, and x^3y product and compute the sums of the xy's, x^2y's, and x^3y's using the Excel 'Sum()' function.

	x	x^2	x^3	x^4	x^5	x^6	y	xy	x^2y	x^3y
	0.0	0.0	0.0	0.0	0.0	0.0	1.0	=R8*X8		
	1.0	1.0	1.0	1.0	1.0	1.0	3.0			
	2.0	4.0	8.0	16.0	32.0	64.0	2.0			
	3.0	9.0	27.0	81.0	243.0	729.0	4.0			
	4.0	16.0	64.0	256.0	1024.0	4096.0	5.0			
Sums =	**10.0**	**30.0**	**100.0**	**354.0**	**1300.0**	**4890.0**	15.0			

Figure 4.17

	x	x^2	x^3	x^4	x^5	x^6	y	xy	x^2y	x^3y
	0.0	0.0	0.0	0.0	0.0	0.0	1.0	0.0	=R8*R8*X8	
	1.0	1.0	1.0	1.0	1.0	1.0	3.0	3.0		
	2.0	4.0	8.0	16.0	32.0	64.0	2.0	4.0		
	3.0	9.0	27.0	81.0	243.0	729.0	4.0	12.0		
	4.0	16.0	64.0	256.0	1024.0	4096.0	5.0	20.0		
Sums =	**10.0**	**30.0**	**100.0**	**354.0**	**1300.0**	**4890.0**	**15.0**			

Figure 4.18

	x	x^2	x^3	x^4	x^5	x^6	y	xy	x^2y	x^3y
	0.0	0.0	0.0	0.0	0.0	0.0	1.0	0.0	0.0	=R8*R8*R8*X8
	1.0	1.0	1.0	1.0	1.0	1.0	3.0	3.0	3.0	
	2.0	4.0	8.0	16.0	32.0	64.0	2.0	4.0	8.0	
	3.0	9.0	27.0	81.0	243.0	729.0	4.0	12.0	36.0	
	4.0	16.0	64.0	256.0	1024.0	4096.0	5.0	20.0	80.0	
Sums =	**10.0**	**30.0**	**100.0**	**354.0**	**1300.0**	**4890.0**	**15.0**			

Figure 4.19

	x	x^2	x^3	x^4	x^5	x^6	y	xy	x^2y	x^3y
	0.0	0.0	0.0	0.0	0.0	0.0	1.0	0.0	0.0	0.0
	1.0	1.0	1.0	1.0	1.0	1.0	3.0	3.0	3.0	3.0
	2.0	4.0	8.0	16.0	32.0	64.0	2.0	4.0	8.0	16.0
	3.0	9.0	27.0	81.0	243.0	729.0	4.0	12.0	36.0	108.0
	4.0	16.0	64.0	256.0	1024.0	4096.0	5.0	20.0	80.0	320.0
Sums =	**10.0**	**30.0**	**100.0**	**354.0**	**1300.0**	**4890.0**	**15.0**	=SUM(Y8:Y12)		

Figure 4.20

	x	x^2	x^3	x^4	x^5	x^6	y	xy	x^2y	x^3y
	0.0	0.0	0.0	0.0	0.0	0.0	1.0	0.0	0.0	0.0
	1.0	1.0	1.0	1.0	1.0	1.0	3.0	3.0	3.0	3.0
	2.0	4.0	8.0	16.0	32.0	64.0	2.0	4.0	8.0	16.0
	3.0	9.0	27.0	81.0	243.0	729.0	4.0	12.0	36.0	108.0
	4.0	16.0	64.0	256.0	1024.0	4096.0	5.0	20.0	80.0	320.0
Sums =	**10.0**	**30.0**	**100.0**	**354.0**	**1300.0**	**4890.0**	**15.0**	**39.0**	=SUM(Z8:Z12)	

Figure 4.21

	x	x^2	x^3	x^4	x^5	x^6	y	xy	x^2y	x^3y
	0.0	0.0	0.0	0.0	0.0	0.0	1.0	0.0	0.0	0.0
	1.0	1.0	1.0	1.0	1.0	1.0	3.0	3.0	3.0	3.0
	2.0	4.0	8.0	16.0	32.0	64.0	2.0	4.0	8.0	16.0
	3.0	9.0	27.0	81.0	243.0	729.0	4.0	12.0	36.0	108.0
	4.0	16.0	64.0	256.0	1024.0	4096.0	5.0	20.0	80.0	320.0
Sums =	**10.0**	**30.0**	**100.0**	**354.0**	**1300.0**	**4890.0**	**15.0**	**39.0**	**127.0**	=SUM(AA8:AA12)

Figure 4.22

	x	x^2	x^3	x^4	x^5	x^6	y	xy	x^2y	x^3y
	0.0	0.0	0.0	0.0	0.0	0.0	1.0	0.0	0.0	0.0
	1.0	1.0	1.0	1.0	1.0	1.0	3.0	3.0	3.0	3.0
	2.0	4.0	8.0	16.0	32.0	64.0	2.0	4.0	8.0	16.0
	3.0	9.0	27.0	81.0	243.0	729.0	4.0	12.0	36.0	108.0
	4.0	16.0	64.0	256.0	1024.0	4096.0	5.0	20.0	80.0	320.0
Sums =	**10.0**	**30.0**	**100.0**	**354.0**	**1300.0**	**4890.0**	**15.0**	**39.0**	**127.0**	**447.0**

Figure 4.23

Step 6. Put the Count() value and the SUM() values into a 4-by-4 (4 rows by 4 columns) and a 4-by-1 (4 rows by 1 column) set of Excel cells as shown in the Excel screen shot below to create the **A** matrix and the **B** vector.

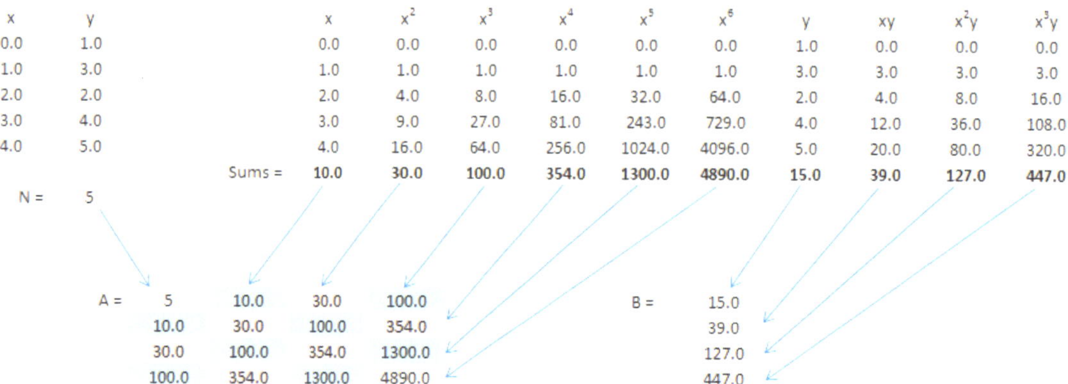

x	y
0.0	1.0
1.0	3.0
2.0	2.0
3.0	4.0
4.0	5.0
N =	5

	x	x^2	x^3	x^4	x^5	x^6	y	xy	x^2y	x^3y
	0.0	0.0	0.0	0.0	0.0	0.0	1.0	0.0	0.0	0.0
	1.0	1.0	1.0	1.0	1.0	1.0	3.0	3.0	3.0	3.0
	2.0	4.0	8.0	16.0	32.0	64.0	2.0	4.0	8.0	16.0
	3.0	9.0	27.0	81.0	243.0	729.0	4.0	12.0	36.0	108.0
	4.0	16.0	64.0	256.0	1024.0	4096.0	5.0	20.0	80.0	320.0
Sums =	**10.0**	**30.0**	**100.0**	**354.0**	**1300.0**	**4890.0**	**15.0**	**39.0**	**127.0**	**447.0**

A =					B =
	5	10.0	30.0	100.0	15.0
	10.0	30.0	100.0	354.0	39.0
	30.0	100.0	354.0	1300.0	127.0
	100.0	354.0	1300.0	4890.0	447.0

Figure 4.24

Step 7. Calculate the inverse of the $\mathbf{A}$ matrix ($\mathbf{A}^{-1}$) by marking off a 4-by-4 array of cells in Excel.

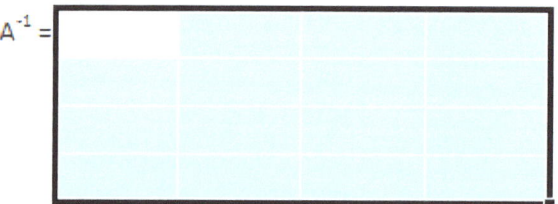

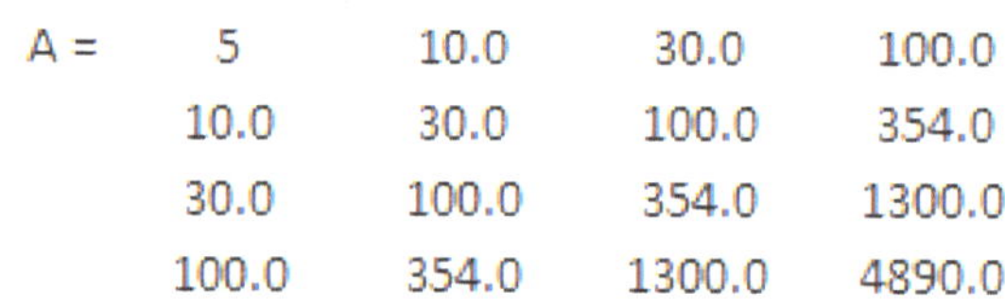

A =				
	5	10.0	30.0	100.0
	10.0	30.0	100.0	354.0
	30.0	100.0	354.0	1300.0
	100.0	354.0	1300.0	4890.0

Figure 4.25

Enter into the first cell of the $\mathbf{A}^{-1}$ matrix the Excel 'MINVERSE()' function. To enter the function, type '=MINVERSE(range)', where 'range' is the range of cells of the $\mathbf{A}$ matrix, then hold down the 'Ctrl' and 'Shift' keys and press 'Enter'.

A =

5	10.0	30.0	100.0
10.0	30.0	100.0	354.0
30.0	100.0	354.0	1300.0
100.0	354.0	1300.0	4890.0

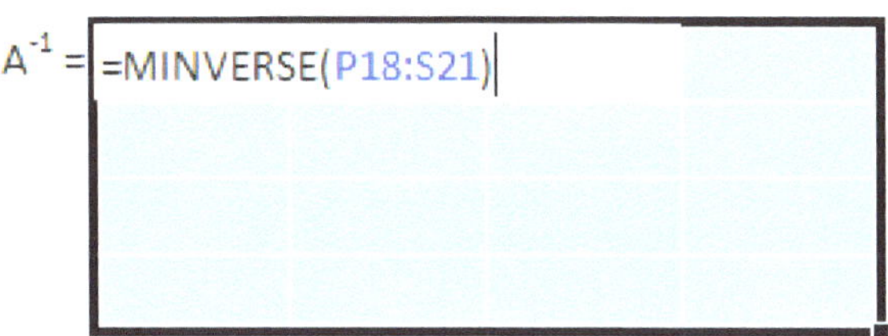

Figure 4.26

Hold down the <Ctrl> and <Shift> keys and press <Enter>.

A =

5	10.0	30.0	100.0
10.0	30.0	100.0	354.0
30.0	100.0	354.0	1300.0
100.0	354.0	1300.0	4890.0

A^{-1} =

0.986	-1.488	0.643	-0.083
-1.488	6.379	-3.869	0.597
0.643	-3.869	2.571	-0.417
-0.083	0.597	-0.417	0.069

Figure 4.27

Step 8. Calculate the coefficients of the cubic least squares curve fit by marking off a 4-by-1 array of cells in Excel.

A =	5	10.0	30.0	100.0		B =	15.0
	10.0	30.0	100.0	354.0			39.0
	30.0	100.0	354.0	1300.0			127.0
	100.0	354.0	1300.0	4890.0			447.0

A^{-1} =	0.986	-1.488	0.643	-0.083		X =		= a_0
	-1.488	6.379	-3.869	0.597				= a_1
	0.643	-3.869	2.571	-0.417				= a_2
	-0.083	0.597	-0.417	0.069				= a_3

Figure 4.28

Enter into the first cell of the **X** vector the Excel 'MMULT()' function. To enter the function, type '=MMULT(range1,range2)', where 'range1' is the range of cells of the $\mathbf{A}^{-1}$ matrix and 'range2' is the range of cells of the **B** vector, then hold down the 'Ctrl' and 'Shift' keys and press 'Enter'.

A =	5	10.0	30.0	100.0		B =	15.0
	10.0	30.0	100.0	354.0			39.0
	30.0	100.0	354.0	1300.0			127.0
	100.0	354.0	1300.0	4890.0			447.0

A^{-1} =	0.986	-1.488	0.643	-0.083	=MMULT(P24:S27,V18:V21)	
	-1.488	6.379	-3.869	0.597		= a_1
	0.643	-3.869	2.571	-0.417		= a_2
	-0.083	0.597	-0.417	0.069		= a_3

Figure 4.29

Hold down the <Ctrl> and <Shift> keys and press <Enter>.

A =	5	10.0	30.0	100.0	B =	15.0	
	10.0	30.0	100.0	354.0		39.0	
	30.0	100.0	354.0	1300.0		127.0	
	100.0	354.0	1300.0	4890.0		447.0	

A^{-1} =	0.986	-1.488	0.643	-0.083	X =	1.143	= a_0
	-1.488	6.379	-3.869	0.597		2.048	= a_1
	0.643	-3.869	2.571	-0.417		-0.929	= a_2
	-0.083	0.597	-0.417	0.069		0.167	= a_3

Figure 4.30

The $\mathbf{X}$ vector contains the coefficients of the best cubic curve fit of the given data of the form $y = a_0 + a_1 x + a_2 x^2 + a_3 x^3$, based on the least squares criterion.

In this example, the best cubic curve fit of the given data, based on the least squares criterion, is $y = 1.143 + 2.048x - 0.929x^2 + 0.167x^3$. This is the cubic equation that minimizes the sum of the squares of the differences for the curve fit.

We can confirm this fit by using Excel to draw a graph of the original data and the cubic curve fit.

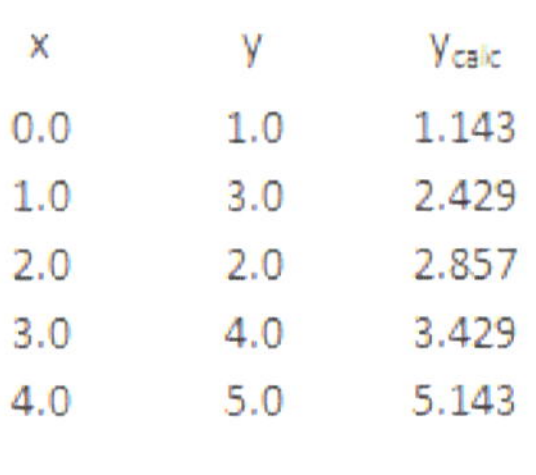

x	y	y_{calc}
0.0	1.0	1.143
1.0	3.0	2.429
2.0	2.0	2.857
3.0	4.0	3.429
4.0	5.0	5.143

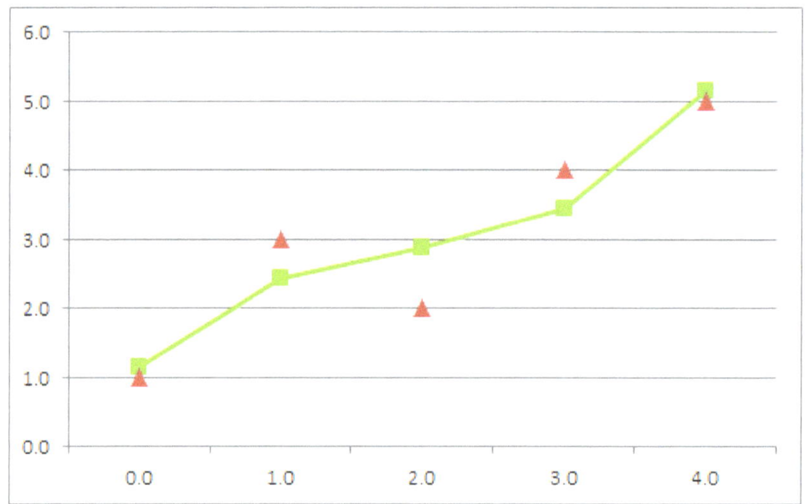

Figure 4.31

Here, the table shows the original x, y data along with the y-values calculated from the cubic curve fit equation (y_{calc}). The triangular points on the graph are the original (5) data points while the squares are the calculated y-values corresponding to the given x-values. Note that the y-intercept of the line is 1.143 which corresponds to the value of a_0 and the value of y when $x = 0$.

Plotting the graph with more calculated points to show the smoothed polynomial curve.

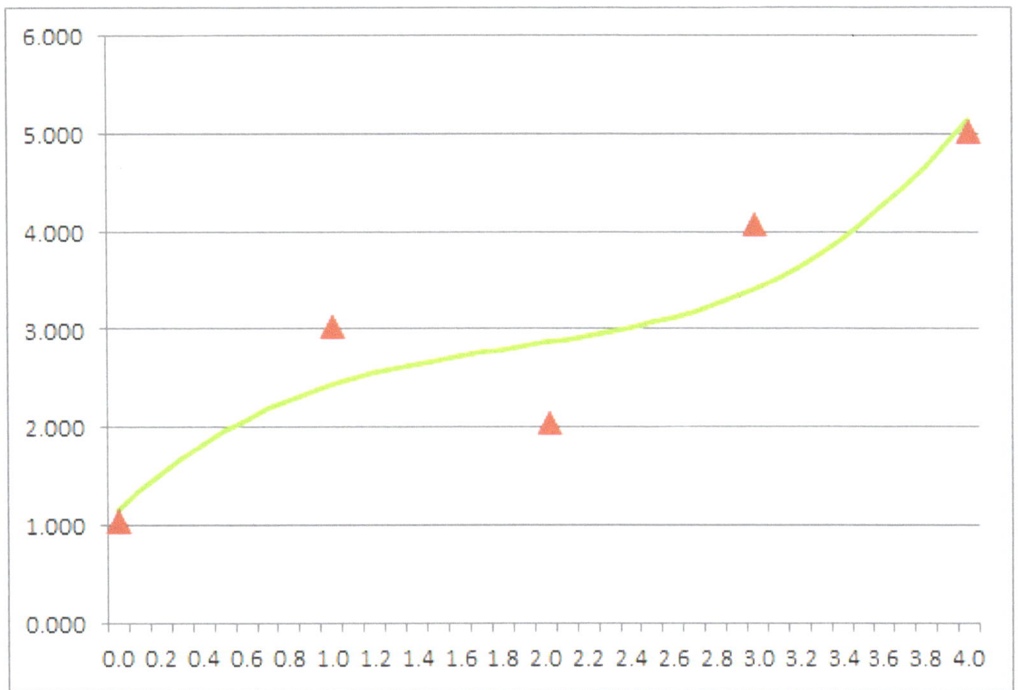

Figure 4.32

CHAPTER 5: HOW TO FIT N POINTS WITH AN (N-1)TH ORDER POLYNOMIAL

WEB VIDEO ADDRESS:
WWW.SONSHINECS.COM/VIDEOS/POLYNOMIAL_VIDEO.MP4

The curve fits described above can easily be extended to include higher-order polynomials used to fit sets of higher numbers of pairs of points. If the number of points to be fitted is n, a least squares fit of a polynomial of order up to $n-2$ can be obtained. For example, a table of 10 pairs of x, y-values could be fitted with least squares polynomial fits of order 1, 2, 3, 4, 5, 6, 7, or 8.

Such a high-ordered polynomial seldom seems justified, however, because one might just as well then find a polynomial of high enough order so as to fit each point exactly, which could be accomplished with an $n-1$ polynomial. That is, for example, we can fit 10 points with a ninth-order polynomial that would pass exactly through each point, thus reproducing the given data points exactly.

To illustrate, suppose we have the same table of x, y-values we have used in the previous sections; that is, suppose we have the table of x, y-values

x	y
0.0	1.0
1.0	3.0
2.0	2.0
3.0	4.0
4.0	5.0

Figure 5.1

Here, the number of points n is 5, so we can fit these points exactly with a 4th-order polynomial of the form $y = a_0 + a_1 x + a_2 x^2 + a_3 x^3 + a_4 x^4$ as follows:

Step 1. To exactly fit each of the 5 given points, we need to find values for the constants $a_0, a_1, a_2, a_3, \text{and } a_4$ so that

$$y_1 = a_0 + a_1 x_1 + a_2 x_1^2 + a_3 x_1^3 + a_4 x_1^4$$

$$y_2 = a_0 + a_1 x_2 + a_2 x_2^2 + a_3 x_2^3 + a_4 x_2^4$$

$$y_3 = a_0 + a_1 x_3 + a_2 x_3^2 + a_3 x_3^3 + a_4 x_3^4$$

$$y_4 = a_0 + a_1 x_4 + a_2 x_4^2 + a_3 x_4^3 + a_4 x_4^4$$

$$y_5 = a_0 + a_1 x_5 + a_2 x_5^2 + a_3 x_5^3 + a_4 x_5^4$$

Figure 5.2

where

$$x_1 = 0, \quad y_1 = 1$$

$$x_2 = 1, \quad y_2 = 3$$

$$x_3 = 2, \quad y_3 = 2$$

$$x_4 = 3, \quad y_4 = 4$$

$$x_5 = 4, \quad y_5 = 5$$

Figure 5.3

so that the equations that need to be solved simultaneously are

$$1 = a_0 + a_1 0 + a_2 0^2 + a_3 0^3 + a_4 0^4$$

$$3 = a_0 + a_1 1 + a_2 1^2 + a_3 1^3 + a_4 1^4$$

$$2 = a_0 + a_1 2 + a_2 2^2 + a_3 2^3 + a_4 2^4$$

$$4 = a_0 + a_1 3 + a_2 3^2 + a_3 3^3 + a_4 3^4$$

$$5 = a_0 + a_1 4 + a_2 4^2 + a_3 4^3 + a_4 4^4$$

Figure 5.4

or

$$1 = 1a_0 + 0a_1 + 0a_2 + 0a_3 + 0a_4$$

$$3 = 1a_0 + 1a_1 + 1a_2 + 1a_3 + 1a_4$$

$$2 = 1a_0 + 2a_1 + 4a_2 + 8a_3 + 16a_4$$

$$4 = 1a_0 + 3a_1 + 9a_2 + 27a_3 + 81a_4$$

$$5 = 1a_0 + 4a_1 + 16a_2 + 64a_3 + 256a_4$$

Figure 5.5

Step 2. We note that this is a set of five linear equations in five unknowns (the unknowns are the a_i values). As such, they can be solved simultaneously using matrix methods and functions contained in Microsoft® Excel. To do this, we set up the matrix of coefficients of the unknowns ($\mathbf{A}$ matrix) and the right-hand-side, constant, vector ($\mathbf{B}$ vector) as shown below.

	1.0	0.0	0.0	0.0	0.0		1.0
	1.0	1.0	1.0	1.0	1.0		3.0
A =	1.0	2.0	4.0	8.0	16.0	B =	2.0
	1.0	3.0	9.0	27.0	81.0		4.0
	1.0	4.0	16.0	64.0	256.0		5.0

Figure 5.6

Step 3. Calculate the inverse of the $\mathbf{A}$ matrix ($\mathbf{A}^{-1}$) by marking off a 5-by-5 array of cells (the same size array as the $\mathbf{A}$ matrix) in Excel.

	1.0	0.0	0.0	0.0	0.0
	1.0	1.0	1.0	1.0	1.0
A =	1.0	2.0	4.0	8.0	16.0
	1.0	3.0	9.0	27.0	81.0
	1.0	4.0	16.0	64.0	256.0

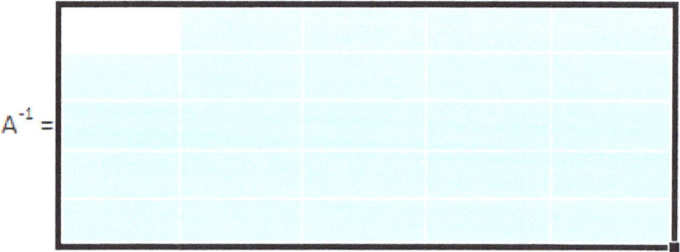

Figure 5.7

Enter into the first cell of the $\mathbf{A}^{-1}$ matrix the Excel 'MINVERSE()' function. To enter the function, type '=MINVERSE(range)', where 'range' is the range of cells of the $\mathbf{A}$ matrix, then hold down the 'Ctrl' and 'Shift' keys and press 'Enter'.

	1.0	0.0	0.0	0.0	0.0
	1.0	1.0	1.0	1.0	1.0
A =	1.0	2.0	4.0	8.0	16.0
	1.0	3.0	9.0	27.0	81.0
	1.0	4.0	16.0	64.0	256.0

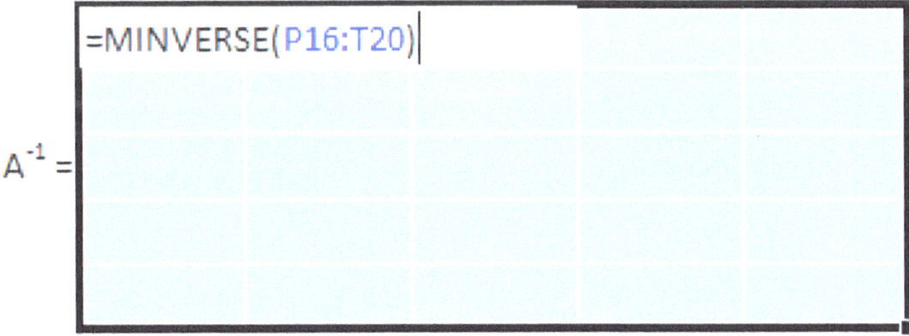

Figure 5.8

Hold down the <Ctrl> and <Shift> keys and press <Enter>.

	1.0	0.0	0.0	0.0	0.0
	1.0	1.0	1.0	1.0	1.0
A =	1.0	2.0	4.0	8.0	16.0
	1.0	3.0	9.0	27.0	81.0
	1.0	4.0	16.0	64.0	256.0

	1.0000	0.0000	0.0000	0.0000	0.0000
	-2.0833	4.0000	-3.0000	1.3333	-0.2500
A^{-1} =	1.4583	-4.3333	4.7500	-2.3333	0.4583
	-0.4167	1.5000	-2.0000	1.1667	-0.2500
	0.0417	-0.1667	0.2500	-0.1667	0.0417

Figure 5.9

Step 4. Calculate the coefficients (the a_i values) of the polynomial curve fit by marking off a 5-by-1 array of cells in Excel.

	1.0	0.0	0.0	0.0	0.0		1.0
	1.0	1.0	1.0	1.0	1.0		3.0
A =	1.0	2.0	4.0	8.0	16.0	B =	2.0
	1.0	3.0	9.0	27.0	81.0		4.0
	1.0	4.0	16.0	64.0	256.0		5.0

	1.0000	0.0000	0.0000	0.0000	0.0000		
	-2.0833	4.0000	-3.0000	1.3333	-0.2500		
A^{-1} =	1.4583	-4.3333	4.7500	-2.3333	0.4583	X =	
	-0.4167	1.5000	-2.0000	1.1667	-0.2500		
	0.0417	-0.1667	0.2500	-0.1667	0.0417		

Figure 5.10

Enter into the first cell of the **X** vector the Excel 'MMULT()' function. To enter the function, type '=MMULT(range1,range2)', where 'range1' is the range of cells of the $\mathbf{A}^{-1}$ matrix and 'range2' is the range of cells of the **B** vector, then hold down the 'Ctrl' and 'Shift' keys and press 'Enter'.

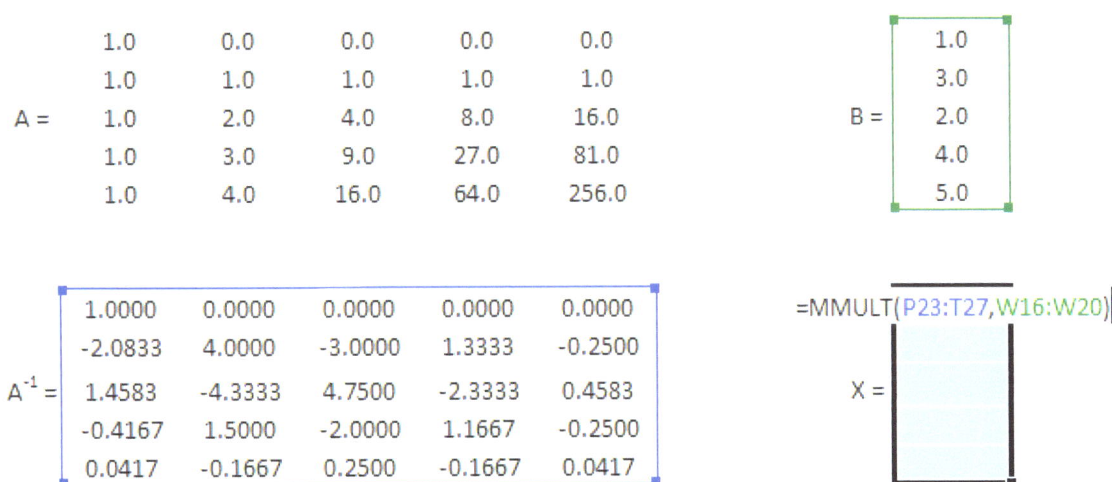

A =

1.0	0.0	0.0	0.0	0.0
1.0	1.0	1.0	1.0	1.0
1.0	2.0	4.0	8.0	16.0
1.0	3.0	9.0	27.0	81.0
1.0	4.0	16.0	64.0	256.0

B =

1.0
3.0
2.0
4.0
5.0

A⁻¹ =

1.0000	0.0000	0.0000	0.0000	0.0000
-2.0833	4.0000	-3.0000	1.3333	-0.2500
1.4583	-4.3333	4.7500	-2.3333	0.4583
-0.4167	1.5000	-2.0000	1.1667	-0.2500
0.0417	-0.1667	0.2500	-0.1667	0.0417

Figure 5.11

Hold down the <Ctrl> and <Shift> keys and press <Enter>.

	1.0	0.0	0.0	0.0	0.0		1.0
	1.0	1.0	1.0	1.0	1.0		3.0
A =	1.0	2.0	4.0	8.0	16.0	B =	2.0
	1.0	3.0	9.0	27.0	81.0		4.0
	1.0	4.0	16.0	64.0	256.0		5.0

	1.0000	0.0000	0.0000	0.0000	0.0000		1.0000	$= a_0$
	-2.0833	4.0000	-3.0000	1.3333	-0.2500		8.0000	$= a_1$
$A^{-1} =$	1.4583	-4.3333	4.7500	-2.3333	0.4583	X =	-9.0833	$= a_2$
	-0.4167	1.5000	-2.0000	1.1667	-0.2500		3.5000	$= a_3$
	0.0417	-0.1667	0.2500	-0.1667	0.0417		-0.4167	$= a_4$

Figure 5.12

The $\mathbf{X}$ vector contains the coefficients of the 4th-order polynomial curve fit of the given data of the form $y = a_0 + a_1 x + a_2 x^2 + a_3 x^3 + a_4 x^4$. In this example, the polynomial curve fit of the given data is

$$y = 1.0000 + 8.0000\,x - 9.0833\,x^2 + 3.5000\,x^3 - 0.4167\,x^4.$$

We can confirm this fit by using Excel to draw a graph of the original data and the polynomial curve fit.

x	y	y_{calc}
0.0	1.0	1.00
1.0	3.0	3.00
2.0	2.0	2.00
3.0	4.0	4.00
4.0	5.0	5.00

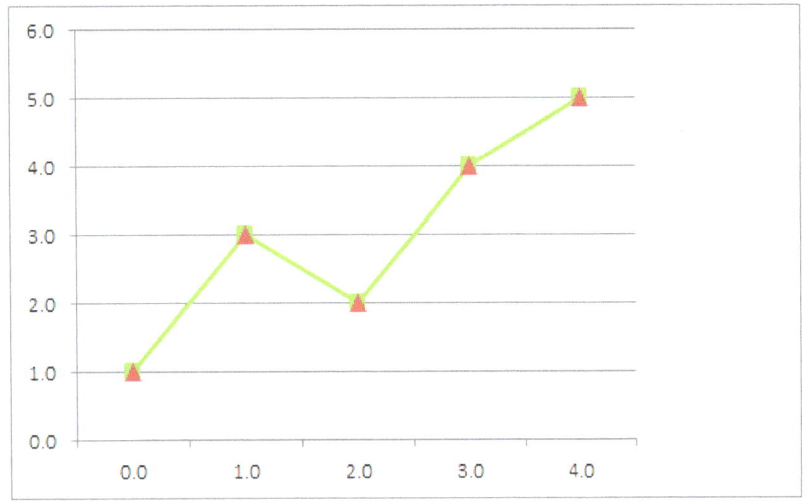

Figure 5.13

Here, the table shows the original x, y data along with the y-values calculated from the 4th-order polynomial curve fit equation (y_{calc}). The triangular points on the graph are the original (5) data points while the squares are the calculated y-values corresponding to the given x-values. Note that the calculated data points match the given data points exactly.

Plotting the graph with more calculated points to show the smoothed polynomial curve:

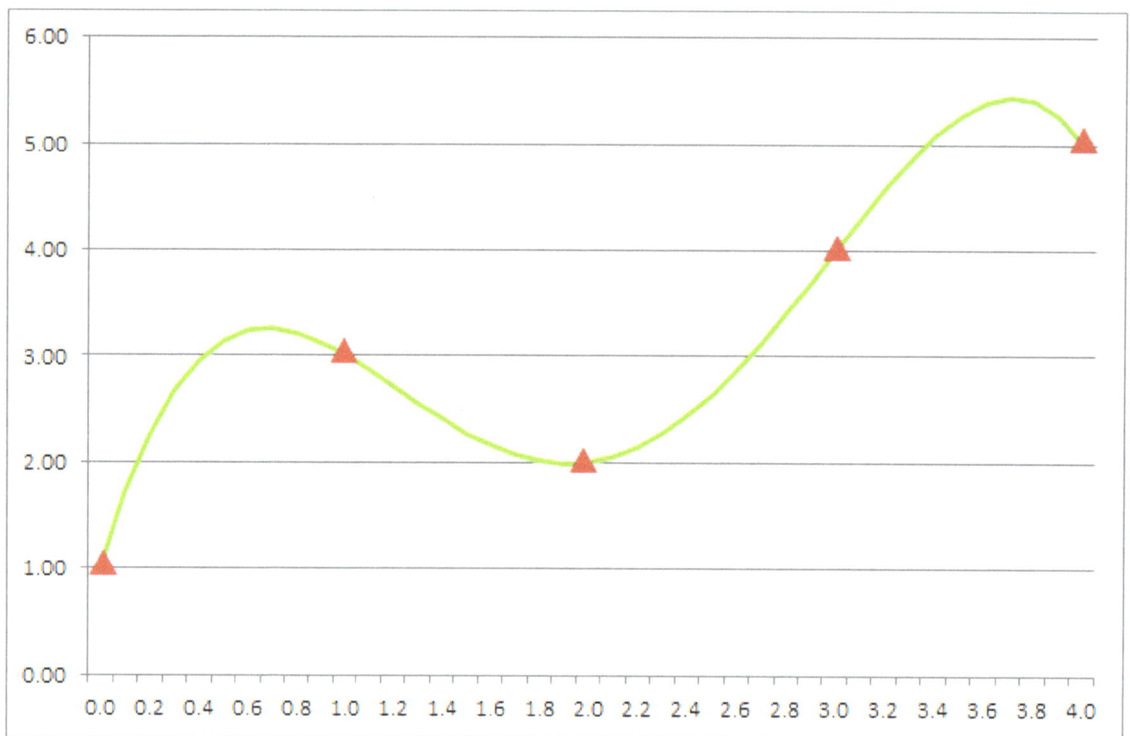
6.00
5.00
4.00
3.00
2.00
1.00
0.00
0.0 0.2 0.4 0.6 0.8 1.0 1.2 1.4 1.6 1.8 2.0 2.2 2.4 2.6 2.8 3.0 3.2 3.4 3.6 3.8 4.0

Figure 5.14

CHAPTER 6: QUALITY OF THE LEAST SQUARES FIT

Regardless of the method used to fit a set of data, it would be nice if there was a way of quantitatively evaluating the quality of the fit. In other words, it would be nice to have a numerical value that would represent how well the analytical equation actually represents the raw data.

Fortunately, the method of least squares has inherent in it the ability to establish such a value. Remember that the least squares method seeks to minimize the sum of the squares of the differences as previously defined. Thus, for any polynomial representation of the data, we can calculate the sum of the squares of the differences for that polynomial representation.

Consider, for example, the fits we have discussed in the previous chapters. The coefficients for the four polynomial fits we have made are given in the table below.

	Linear	Quadratic	Cubic	Quartic
a_0 =	1.2000	1.3430	1.1430	1.0000
a_1 =	0.9000	0.6140	2.0480	8.0000
a_2 =		0.0710	-0.9290	-9.0833
a_3 =			0.1670	3.5000
a_4 =				-0.4167

Figure 6.1

From the coefficients and the raw data, we can calculate the sum of the squares of the differences for each of the polynomial fits. These sums are shown in the following table:

		y_{calc}				Differences Squared			
x	y	Linear	Quadratic	Cubic	Quartic	Linear	Quadratic	Cubic	Quartic
0.0	1.0	1.2000	1.3430	1.1430	1.0000	0.0400	0.1176	0.0204	0.0000
1.0	3.0	2.1000	2.0280	2.4290	3.0000	0.8100	0.9448	0.3260	0.0000
2.0	2.0	3.0000	2.8550	2.8590	1.9996	1.0000	0.7310	0.7379	0.0000
3.0	4.0	3.9000	3.8240	3.4350	3.9976	0.0100	0.0310	0.3192	0.0000
4.0	5.0	4.8000	4.9350	5.1590	4.9920	0.0400	0.0042	0.0253	0.0001
					Sums =	1.9000	1.8287	1.4289	0.0001

Figure 6.2

Note that, as the order of the polynomial fit increases, the sum of the squares of the differences decreases until, in the limit, when the polynomial reproduces the data points exactly, the sum of the squares of the differences is zero.

We should expect that, for any given set of raw data, the sum of the squares of the differences decreases as the order of the polynomial fitting those data increases since we should expect that higher and higher ordered polynomials will come closer and closer to representing the raw data exactly. This phenomenon is exhibited by the table above.

When we have two different sets of raw data, each with the same number of data points, both of which are fitted with, say, a quadratic least squares equation, to find which of the two sets of data are better fitted by the quadratic, we can look at the sum of the squares of the differences of the two fits. The one with the smaller sum of the squares would be the better fit, based on the least squares criterion.

One caveat to be noted when comparing values of the sum of the squares of the differences is that this numerical value will be affected by the number of data points that are being fitted. The more data points being fitted, the more sums of the squares make up the total sum of the squares so that the total numerical value will tend to be larger simply because of the number of values being summed. Thus, in cases where the sum of the squares of the differences is being compared for a curve fit of a larger number of data points, it may be more meaningful to compare the sum of the squares of the differences per data point; that is, compare the sum of the squares of the differences divided by the number of data points being fitted.

CHAPTER 7: APPLICATIONS

CHECK PROCESSING

WEB VIDEO ADDRESS:

WWW.SONSHINECS.COM/VIDEOS/CHECK_PROCESSING_VIDEO.MP4

A bank has recorded data showing the number of checks processed by the bank in the past twenty-four (24) months (see the table below). The bank manager would like to know the number of checks they can be expected to process each month during the next six (6) months. In addition, the bank's current processing system (equipment, personnel, etc.) is limited to being able to process a maximum of 175,000 checks per month, so the manager would like to know approximately when the bank will reach that processing level so he can plan for the expenditures needed to upgrade to a new or enhanced check processing system.

Month	Number of Checks In Thousands	Month	Number of Checks In Thousands
1	52	13	72
2	55	14	78
3	58	15	84
4	62	16	80

5	60	17	88
6	57	18	96
7	63	19	110
8	65	20	105
9	62	21	115
10	70	22	130
11	75	23	135
12	77	24	140

Table 7.1.1

To address the bank manager's questions, we will use the method of least squares to fit the given data with linear, quadratic, and cubic polynomial equations. We will use Microsoft® Excel for the calculations.

To begin, we enter the data into a table in Excel and create a graph of the raw data as shown in the figure below.

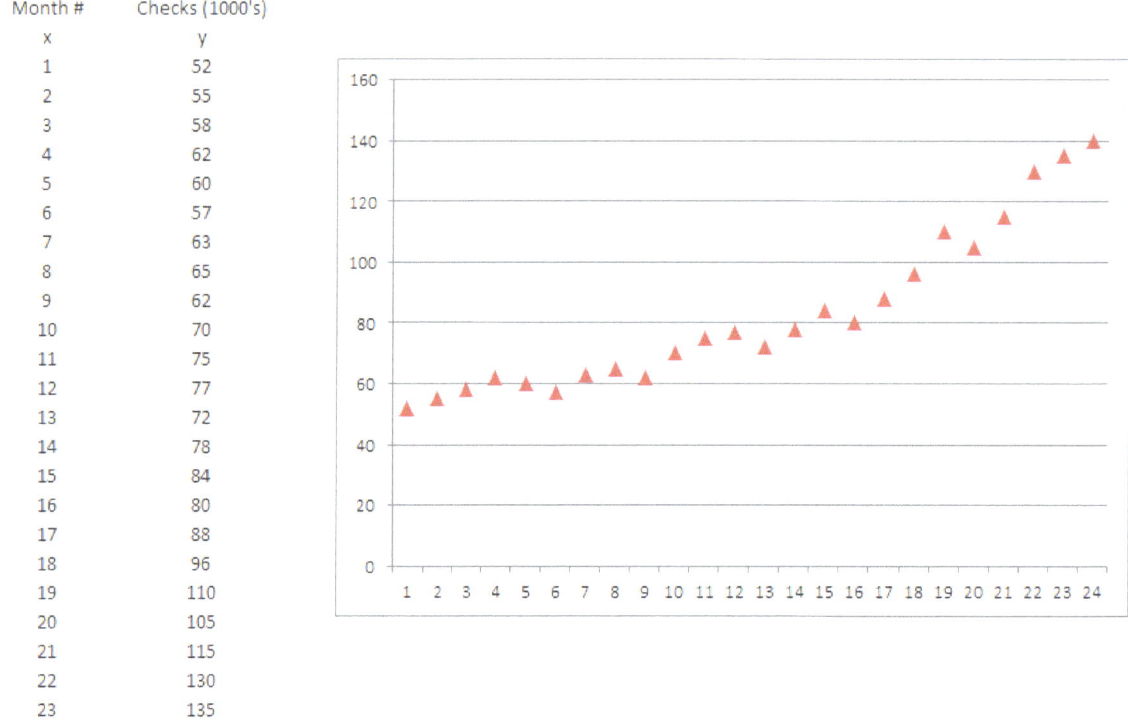

Month #	Checks (1000's)
x	y
1	52
2	55
3	58
4	62
5	60
6	57
7	63
8	65
9	62
10	70
11	75
12	77
13	72
14	78
15	84
16	80
17	88
18	96
19	110
20	105
21	115
22	130
23	135
24	140

Figure 7.1.1

We calculate the coefficients of the linear least squares curve fit of the data of the form $y = a_0 + a_1 x$ in Excel as in Figure 7.1.2.

	x	x^2	y	xy
	1	1	52	52
	2	4	55	110
	3	9	58	174
	4	16	62	248
	5	25	60	300
	6	36	57	342
	7	49	63	441
	8	64	65	520
	9	81	62	558
	10	100	70	700
	11	121	75	825
	12	144	77	924
	13	169	72	936
	14	196	78	1092
	15	225	84	1260
	16	256	80	1280
	17	289	88	1496
	18	324	96	1728
	19	361	110	2090
	20	400	105	2100
	21	441	115	2415
	22	484	130	2860
	23	529	135	3105
	24	576	140	3360
Sums =	300	4900	1989	28916

A =	24	300	B =	1989	
	300	4900		28916	
A^{-1} =	0.1775	-0.0109	X =	38.8152	$= a_0$
	-0.0109	0.0009		3.5248	$= a_1$

Figure 7.1.2

So the best linear, straight-line polynomial to represent the data based on the least squares criterion is $y = 38.8152 + 3.5248x$.

Similarly, we calculate the coefficients of the quadratic least squares curve fit of the data of the form $y = a_0 + a_1 x + a_2 x^2$ in Excel as in Figure 7.1.3.

	x	x^2	x^3	x^4	y	xy	x^2y
	1	1	1	1	52	52	52
	2	4	8	16	55	110	220
	3	9	27	81	58	174	522
	4	16	64	256	62	248	992
	5	25	125	625	60	300	1500
	6	36	216	1296	57	342	2052
	7	49	343	2401	63	441	3087
	8	64	512	4096	65	520	4160
	9	81	729	6561	62	558	5022
	10	100	1,000	10000	70	700	7000
	11	121	1,331	14641	75	825	9075
	12	144	1,728	20736	77	924	11088
	13	169	2,197	28561	72	936	12168
	14	196	2,744	38416	78	1092	15288
	15	225	3,375	50625	84	1260	18900
	16	256	4,096	65536	80	1280	20480
	17	289	4,913	83521	88	1496	25432
	18	324	5,832	104976	96	1728	31104
	19	361	6,859	130321	110	2090	39710
	20	400	8,000	160000	105	2100	42000
	21	441	9,261	194481	115	2415	50715
	22	484	10,648	234256	130	2860	62920
	23	529	12,167	279841	135	3105	71415
	24	576	13,824	331776	140	3360	80640
Sums =	300	4900	90000	1763020	1989	28916	515542

	24	300	4900		1989	
A =	300	4900	90000	B =	28916	
	4900	90000	1763020		515542	
	0.4452	-0.0726	0.0025		58.867	$= a_0$
$A^{-1} =$	-0.0726	0.0151	-0.0006	X =	-1.103	$= a_1$
	0.0025	-0.0006	0.0000		0.185	$= a_2$

Figure 7.1.3

From which we get the best quadratic equation to represent the data based on the least squares criterion is $y = 58.867 - 1.103x + 0.185x^2$.

We use Excel to calculate the coefficients of the cubic least squares curve fit of the data of the form $y = a_0 + a_1 x + a_2 x^2 + a_3 x^3$ as in Figure 7.1.4 below.

	x	x^2	x^3	x^4	x^5	x^6	y	xy	x^2y	x^3y
	1	1	1	1	1	1	52	52	52	52
	2	4	8	16	32	64	55	110	220	440
	3	9	27	81	243	729	58	174	522	1566
	4	16	64	256	1,024	4096	62	248	992	3968
	5	25	125	625	3,125	15625	60	300	1500	7500
	6	36	216	1296	7,776	46656	57	342	2052	12312
	7	49	343	2401	16,807	117649	63	441	3087	21609
	8	64	512	4096	32,768	262144	65	520	4160	33280
	9	81	729	6561	59,049	531441	62	558	5022	45198
	10	100	1,000	10000	100,000	1000000	70	700	7000	70000
	11	121	1,331	14641	161,051	1771561	75	825	9075	99825
	12	144	1,728	20736	248,832	2985984	77	924	11088	133056
	13	169	2,197	28561	371,293	4826809	72	936	12168	158184
	14	196	2,744	38416	537,824	7529536	78	1092	15288	214032
	15	225	3,375	50625	759,375	11390625	84	1260	18900	283500
	16	256	4,096	65536	1,048,576	16777216	80	1280	20480	327680
	17	289	4,913	83521	1,419,857	24137569	88	1496	25432	432344
	18	324	5,832	104976	1,889,568	34012224	96	1728	31104	559872
	19	361	6,859	130321	2,476,099	47045881	110	2090	39710	754490
	20	400	8,000	160000	3,200,000	64000000	105	2100	42000	840000
	21	441	9,261	194481	4,084,101	85766121	115	2415	50715	1065015
	22	484	10,648	234256	5,153,632	1.13E+08	130	2860	62920	1384240
	23	529	12,167	279841	6,436,343	1.48E+08	135	3105	71415	1642545
	24	576	13,824	331776	7,962,624	1.91E+08	140	3360	80640	1935360
Sums =	300	4900	90000	1763020	35970000	7.55E+08	1989	28916	515542	10026068

	24	300	4900	90000		1989
A =	300	4900	90000	1763020	B =	28916
	4900	90000	1763020	35970000		515542
	90000	1763020	35970000	754740700		10026068

	0.9269	-0.2827	0.0231	-0.000548968		51.155	$= a_0$
A^{-1} =	-0.2827	0.1067	-0.0095	0.000239419	X =	2.261	$= a_1$
	0.0231	-0.0095	0.0009	-2.34602E-05		-0.144	$= a_2$
	-0.00055	0.000239	-2.3E-05	6.25604E-07		0.008789	$= a_3$

Figure 7.1.4

From which we get that the best cubic equation to represent the data based on the least squares criterion is $y = 51.155 + 2.261x - 0.144x^2 + 0.008789x^3$.

The representation of the given data by the three curve fits (linear, quadratic, and cubic) can be illustrated with the following graph (Figure 7.1.5).

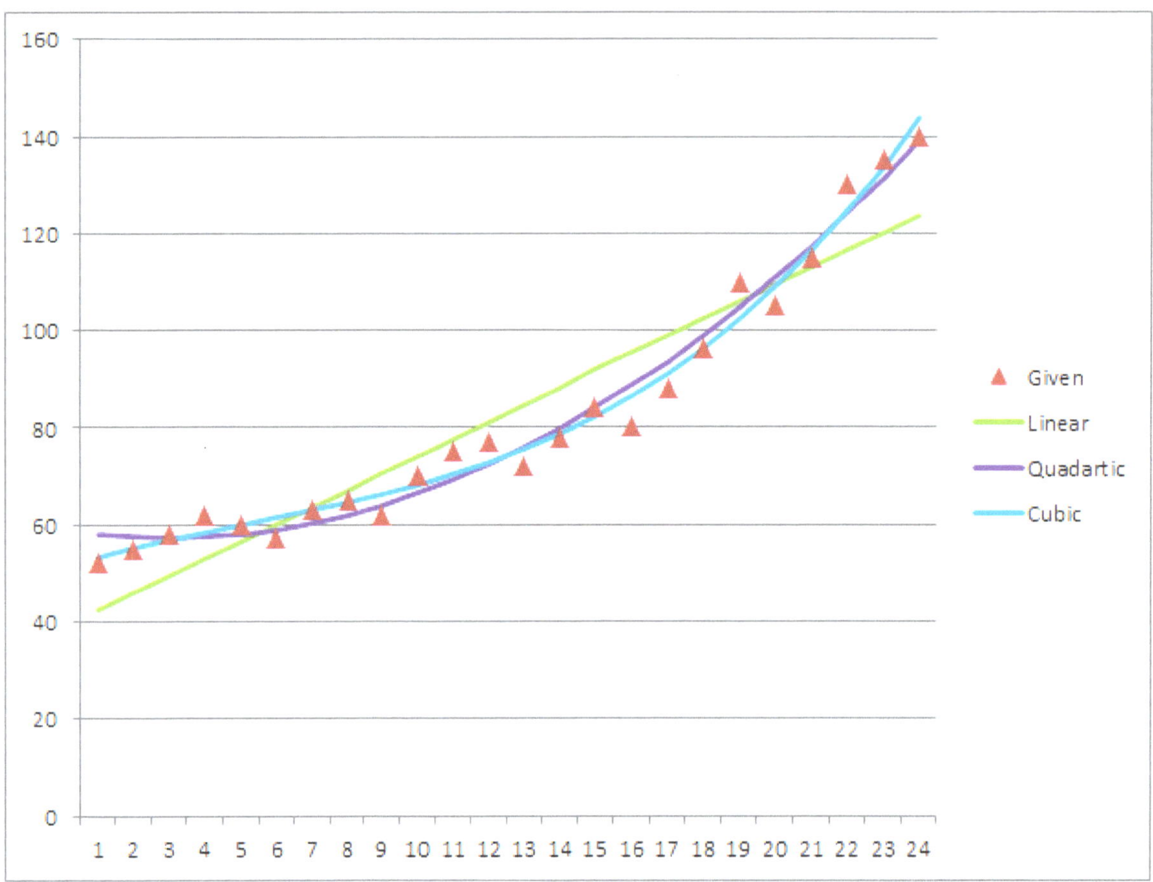

Figure 7.1.5

Now, we want to use the three least squares equations to project or forecast six months into the future. Another way of stating this is to say that we want to use the three equations, linear, quadratic, and cubic, to extrapolate the given data to calculate the number of checks to be processed per month for months 25 through 30; that is, for the six months following the 24 months for which we have check processing data. This forecast calculation, performed using Excel with the mathematical models derived above, is shown below.

Month	Checks (1000's)		
	y_{linear}	$y_{quadratic}$	y_{cubic}
25	126.935	146.987	154.699
26	130.460	155.324	166.738
27	133.984	164.031	179.859
28	137.509	173.109	194.115
29	141.034	182.557	209.559
30	144.559	192.375	226.243

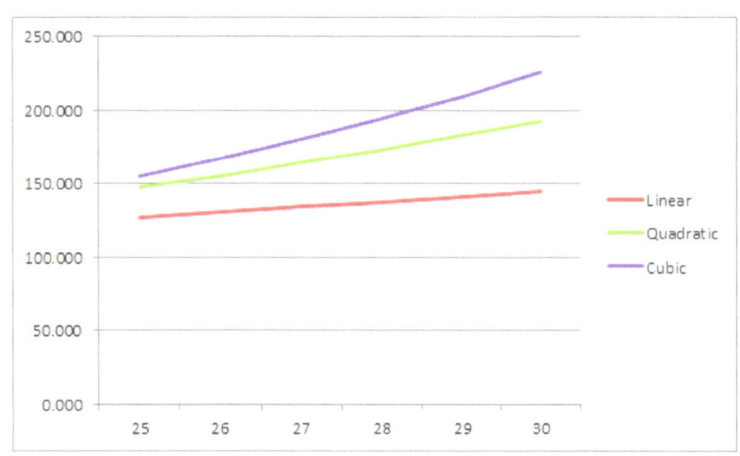

Figure 7.1.6

So, based on a least squares criterion for representing the given data and using those representations to extrapolate or forecast for the next six months, the bank can expect to process between approximately 144,000 and 226,000 checks per month by the sixth month.

With regard to the bank manager's second question about when the bank's checking processing system will reach its maximum capacity, if we assume the worst case scenario for the bank's check processing requirements (that is, the criterion that indicates the largest number of checks to be processed per month, or that the current system's capacity will be reached sooner rather than later), we use the cubic least squares fit of the data to find the month number (x) when the number of checks to be processed (y) will reach a value of 175 thousand.

To do this in Excel, we will use the Solver add-in for Excel. We first designate a specific cell in Excel to contain the value for the variable x, the month number when the check processing rate will reach 175,000. We can also enter some arbitrary number in this cell (cell R13) as in Figure 7.1.7.

	Cubic
a_0 (cell R7) =	51.1548
a_1 (cell R8) =	2.2609
a_2 (cell R9) =	-0.1445
a_3 (cell R10) =	0.0088
Variable x (cell R13):	24.0000
Objective (cell R15):	

Figure 7.1.7

In the 'Objective' cell (cell R15), we enter the formula for the cubic curve fit equation defining the number of checks processed per month, using the 'Variable x:' cell value for the x value.

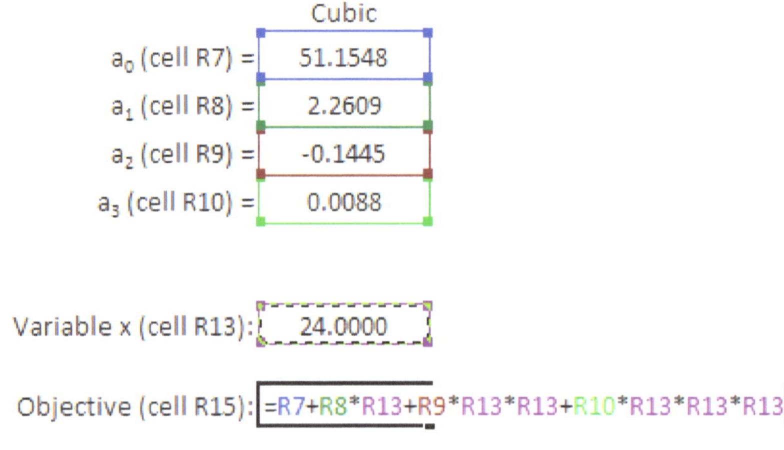

	Cubic
a_0 (cell R7) =	51.1548
a_1 (cell R8) =	2.2609
a_2 (cell R9) =	-0.1445
a_3 (cell R10) =	0.0088
Variable x (cell R13):	24.0000
Objective (cell R15):	=R7+R8*R13+R9*R13*R13+R10*R13*R13*R13

Figure 7.1.8

We now invoke the Solver via the Data…Analysis menu item in Excel and set the Solver Parameters as shown in the figure (Figure 7.1.10).

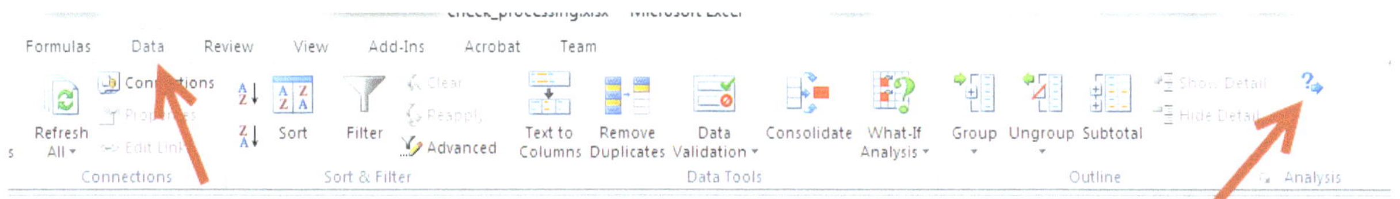

Figure 7.1.9

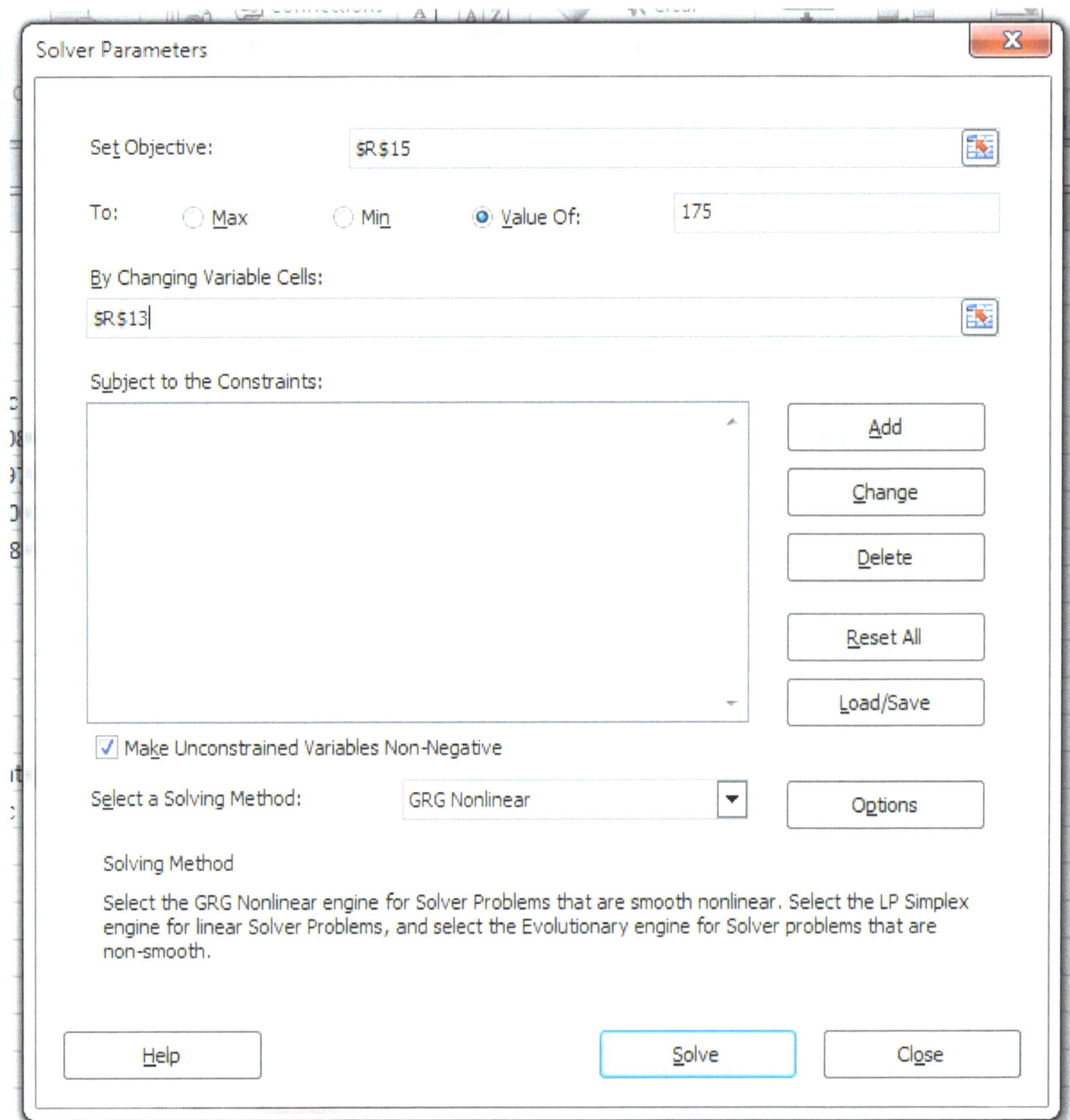

Figure 7.1.10

Click the Solve button.

	Cubic
a_0 (cell R7) =	51.1548
a_1 (cell R8) =	2.2609
a_2 (cell R9) =	-0.1445
a_3 (cell R10) =	0.0088
Variable x (cell R13):	26.6394
Objective (cell R15):	174.9999

Figure 7.1.11

And the Solver calculation tells us that the bank's check processing system with reach its maximum capacity of 175,000 checks approximately two and a half months (roughly 79 days) from the end of the original 24 month period upon which the calculations are based.

STOCK INVESTING

WEB VIDEO ADDRESS:

WWW.SONSHINECS.COM/VIDEOS/STOCK_INVESTING_VIDEO.MP4

A stock market investor is interested in a particular common stock listed on the New York Stock Exchange. Although he does not currently own this particular stock, he has recorded data showing the average weekly closing price of the stock over the past fifty-two (52) weeks (see the table below). The investor would like to know when he should buy (or should have bought) or sold the stock based on an analysis of the price data he has recorded. He proposes to curve fit the price data with a cubic least squares polynomial equation in the hopes that the curve fit will give him an indication of when the stock price is/was at a low point in its weekly price variation, indicating that the stock should be purchased, as well as an indication of when the stock price is/was at a high point in its weekly price variation, indicating that the stock should be sold. (This investor has adopted the strategy of making money in the stock market by buying a stock at a low price and selling it at a higher price, unlike most amateur investors who typically buy at a high price and sell at a lower price.) The buy-low-sell-high strategy is the reason why he wants to use a cubic polynomial fit of the data (as opposed to a linear or quadratic least squares fit), since a cubic equation can exhibit both a high/maximum and a low/minimum point in its curve.

Month	Stock Price, $	Month	Stock Price, $
1	25.25	27	24.30
2	25.00	28	25.10
3	23.90	29	23.75
4	24.00	30	25.75
5	23.30	31	25.60
6	22.25	32	24.15
7	21.95	33	25.00
8	22.15	34	26.05
9	21.60	35	25.10
10	22.00	36	25.20
11	21.60	37	27.50
12	22.10	38	27.80
13	20.00	39	29.10
14	19.90	40	29.80

15	19.55	41	30.00
16	19.20	42	29.10
17	19.75	43	28.10
18	20.10	44	28.25
19	20.50	45	27.85
20	21.90	46	26.35
21	21.15	47	25.30
22	25.55	48	27.95
23	23.25	49	28.15
24	24.35	50	29.10
25	24.05	51	28.15
26	24.45	52	28.00

Table 7.2.1

To address his questions, the investor will use the method of least squares to fit the given stock price data with a cubic polynomial equation. He will use Microsoft® Excel for the calculations.

To begin, he enters the data into a table in Excel and creates a graph of the raw data as shown in the figures below:

Month	Avg. Price, $	Month	Avg. Price, $
1	25.25	27	24.30
2	25.00	28	25.10
3	23.90	29	23.75
4	24.00	30	25.75
5	23.30	31	25.60
6	22.25	32	24.15
7	21.95	33	25.00
8	22.15	34	26.05
9	21.60	35	25.10
10	22.00	36	25.20
11	21.60	37	27.50
12	22.10	38	27.80
13	20.00	39	29.10
14	19.90	40	29.80
15	19.55	41	30.00
16	19.20	42	29.10
17	19.75	43	28.10
18	20.10	44	28.25
19	20.50	45	27.85
20	21.90	46	26.35
21	21.15	47	25.30
22	25.55	48	27.95
23	23.25	49	28.15
24	24.35	50	29.10
25	24.05	51	28.15
26	24.45	52	28.00

Figure 7.2.1

Figure 7.2.2

He proceeds to make the cubic least squares fit of the data in Excel as shown in the following figures:

x	x^2	x^3	x^4	x^5	x^6	y	xy	x^2y	x^3y
1	1	1	1	1	1	25.25	25.25	25.25	25.25
2	4	8	16	32	64	25.00	50.00	100.00	200.00
3	9	27	81	243	729	23.90	71.70	215.10	645.30
4	16	64	256	1024	4096	24.00	96.00	384.00	1536.00
5	25	125	625	3125	15625	23.30	116.50	582.50	2912.50
6	36	216	1296	7776	46656	22.25	133.50	801.00	4806.00
7	49	343	2401	16807	117649	21.95	153.65	1075.55	7528.85
8	64	512	4096	32768	262144	22.15	177.20	1417.60	11340.80
9	81	729	6561	59049	531441	21.60	194.40	1749.60	15746.40
10	100	1000	10000	100000	1000000	22.00	220.00	2200.00	22000.00
11	121	1331	14641	161051	1771561	21.60	237.60	2613.60	28749.60
12	144	1728	20736	248832	2985984	22.10	265.20	3182.40	38188.80
13	169	2197	28561	371293	4826809	20.00	260.00	3380.00	43940.00
14	196	2744	38416	537824	7529536	19.90	278.60	3900.40	54605.60
15	225	3375	50625	759375	11390625	19.55	293.25	4398.75	65981.25
16	256	4096	65536	1048576	16777216	19.20	307.20	4915.20	78643.20
17	289	4913	83521	1419857	24137569	19.75	335.75	5707.75	97031.75
18	324	5832	104976	1889568	34012224	20.10	361.80	6512.40	117223.20
19	361	6859	130321	2476099	47045881	20.50	389.50	7400.50	140609.50
20	400	8000	160000	3200000	64000000	21.90	438.00	8760.00	175200.00
21	441	9261	194481	4084101	85766121	21.15	444.15	9327.15	195870.15
22	484	10648	234256	5153632	113379904	25.55	562.10	12366.20	272056.40
23	529	12167	279841	6436343	148035889	23.25	534.75	12299.25	282882.75
24	576	13824	331776	7962624	191102976	24.35	584.40	14025.60	336614.40
25	625	15625	390625	9765625	244140625	24.05	601.25	15031.25	375781.25
26	676	17576	456976	11881376	308915776	24.45	635.70	16528.20	429733.20

Figure 7.2.3

	27	729	19683	531441	14348907	387420489	24.30	656.10	17714.70	478296.90
	28	784	21952	614656	17210368	481890304	25.10	702.80	19678.40	550995.20
	29	841	24389	707281	20511149	594823321	23.75	688.75	19973.75	579238.75
	30	900	27000	810000	24300000	729000000	25.75	772.50	23175.00	695250.00
	31	961	29791	923521	28629151	887503681	25.60	793.60	24601.60	762649.60
	32	1024	32768	1048576	33554432	1073741824	24.15	772.80	24729.60	791347.20
	33	1089	35937	1185921	39135393	1291467969	25.00	825.00	27225.00	898425.00
	34	1156	39304	1336336	45435424	1544804416	26.05	885.70	30113.80	1023869.20
	35	1225	42875	1500625	52521875	1838265625	25.10	878.50	30747.50	1076162.50
	36	1296	46656	1679616	60466176	2176782336	25.20	907.20	32659.20	1175731.20
	37	1369	50653	1874161	69343957	2565726409	27.50	1017.50	37647.50	1392957.50
	38	1444	54872	2085136	79235168	3010936384	27.80	1056.40	40143.20	1525441.60
	39	1521	59319	2313441	90224199	3518743761	29.10	1134.90	44261.10	1726182.90
	40	1600	64000	2560000	102400000	4096000000	29.80	1192.00	47680.00	1907200.00
	41	1681	68921	2825761	115856201	4750104241	30.00	1230.00	50430.00	2067630.00
	42	1764	74088	3111696	130691232	5489031744	29.10	1222.20	51332.40	2155960.80
	43	1849	79507	3418801	147008443	6321363049	28.10	1208.30	51956.90	2234146.70
	44	1936	85184	3748096	164916224	7256313856	28.25	1243.00	54692.00	2406448.00
	45	2025	91125	4100625	184528125	8303765625	27.85	1253.25	56396.25	2537831.25
	46	2116	97336	4477456	205962976	9474296896	26.35	1212.10	55756.60	2564803.60
	47	2209	103823	4879681	229345007	10779215329	25.30	1189.10	55887.70	2626721.90
	48	2304	110592	5308416	254803968	12230590464	27.95	1341.60	64396.80	3091046.40
	49	2401	117649	5764801	282475249	13841287201	28.15	1379.35	67588.15	3311819.35
	50	2500	125000	6250000	312500000	15625000000	29.10	1455.00	72750.00	3637500.00
	51	2601	132651	6765201	345025251	17596287801	28.15	1435.65	73218.15	3734125.65
	52	2704	140608	7311616	380204032	19770609664	28.00	1456.00	75712.00	3937024.00
Sums =	1378	48230	1898884	79743482	3488249908	156942769490	1279.30	35676.75	1289366.55	51688657.35

Figure 7.2.4

	52	1378	48230	1898884		1279.30
A =	1378	48230	1898884	79743482	B =	35676.75
	48230	1898884	79743482	3488249908		1289366.55
	1898884	79743482	3488249908	156942769490		51688657.35

	0.3567	-0.0509	0.0019	-2.15471E-05		26.453	$= a_0$
A^{-1} =	-0.0509	0.0093	-0.0004	4.6582E-06	X =	-0.910	$= a_1$
	0.0019	-0.0004	0.0000	-2.17647E-07		0.043	$= a_2$
	-2.2E-05	4.66E-06	-2.176E-07	2.7377E-09		-0.00049477	$= a_3$

Figure 7.2.5

So the cubic least squares fit of the given data is $y = 26.453 - 0.910x + 0.043x^2 - 0.0004948x^3$. Using this equation, the investor can set up the following table showing the given stock prices along with the stock prices calculated using the cubic equation:

Month	Avg. Price, $	Avg. Price, calc, $	Month	Avg. Price, $	Avg. Price, calc, $
1	25.25	25.59	27	24.30	23.85
2	25.00	24.80	28	25.10	24.21
3	23.90	24.10	29	23.75	24.57
4	24.00	23.48	30	25.75	24.93
5	23.30	22.93	31	25.60	25.29
6	22.25	22.45	32	24.15	25.65
7	21.95	22.05	33	25.00	26.00
8	22.15	21.71	34	26.05	26.33
9	21.60	21.43	35	25.10	26.66
10	22.00	21.21	36	25.20	26.96
11	21.60	21.05	37	27.50	27.25
12	22.10	20.94	38	27.80	27.51
13	20.00	20.89	39	29.10	27.75
14	19.90	20.88	40	29.80	27.96
15	19.55	20.92	41	30.00	28.14
16	19.20	21.00	42	29.10	28.28
17	19.75	21.12	43	28.10	28.38
18	20.10	21.28	44	28.25	28.45
19	20.50	21.47	45	27.85	28.47
20	21.90	21.69	46	26.35	28.44
21	21.15	21.94	47	25.30	28.36
22	25.55	22.21	48	27.95	28.24
23	23.25	22.51	49	28.15	28.05
24	24.35	22.82	50	29.10	27.81
25	24.05	23.15	51	28.15	27.50
26	24.45	23.49	52	28.00	27.13

Figure 7.2.6

From which the following graph can be generated:

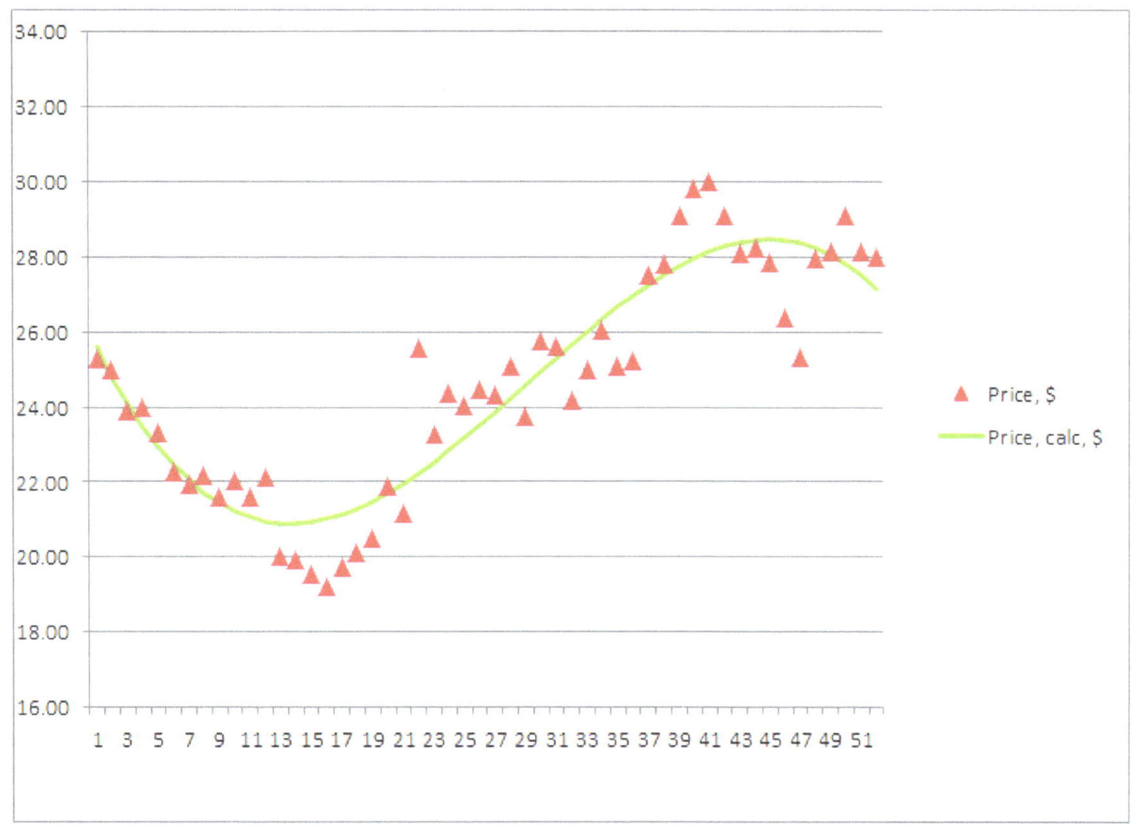

Figure 7.2.7

From the graph of the cubic least squares curve fit, the investor sees that this stock had an indicated buy point at around week 13 and an indicated sell point at around week 45. To identify the buy-sell points more precisely, he uses the Excel Solver add-in as follows:

1. Set up the cubic equation calculation in the Objective Cell, referring to the Variable Cell, to find the minimum parameters of the curve fit.

Curve Fits

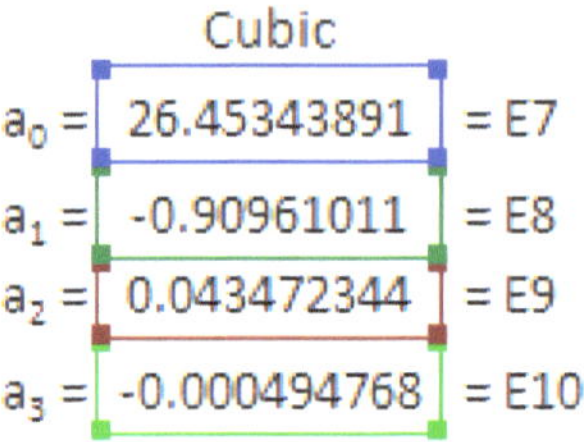

Buy - Minimum Price

Month #: 13.000 = E17

Price, $: =E7+E8*E17+E9*E17*E17+E10*E17*E17*E17

Figure 7.2.8

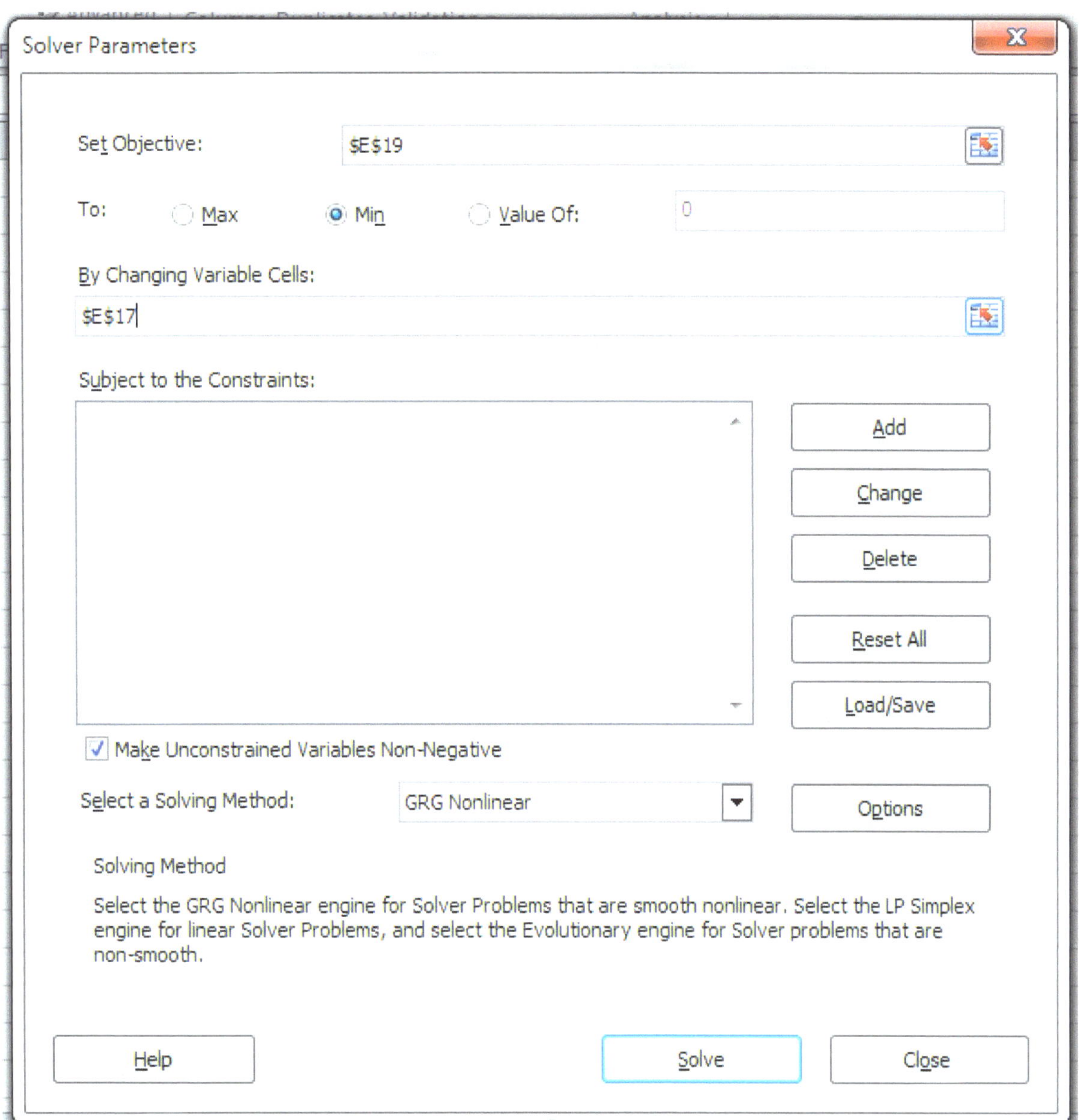

Figure 7.2.9

Click the Solve button.

Buy - Minimum Price

Month #:	13.637	= E17
Price, $:	20.879	= E19

Figure 7.2.10

2. Set up the cubic equation calculation in the Objective Cell, referring to the Variable Cell, to find the maximum parameters of the curve fit.

Curve Fits

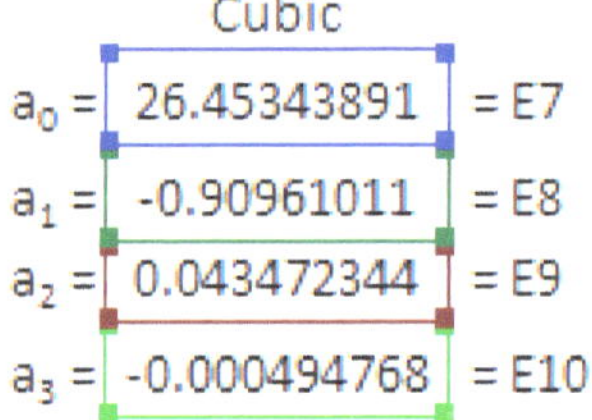

Cubic
a_0 = 26.45343891 = E7
a_1 = -0.90961011 = E8
a_2 = 0.043472344 = E9
a_3 = -0.000494768 = E10

Sell - Maximum Price

Month #:	45.000	= E26
Price, $:	=E7+E8*E26+E9*E26*E26+E10*E26*E26*E26	

Figure 7.2.11

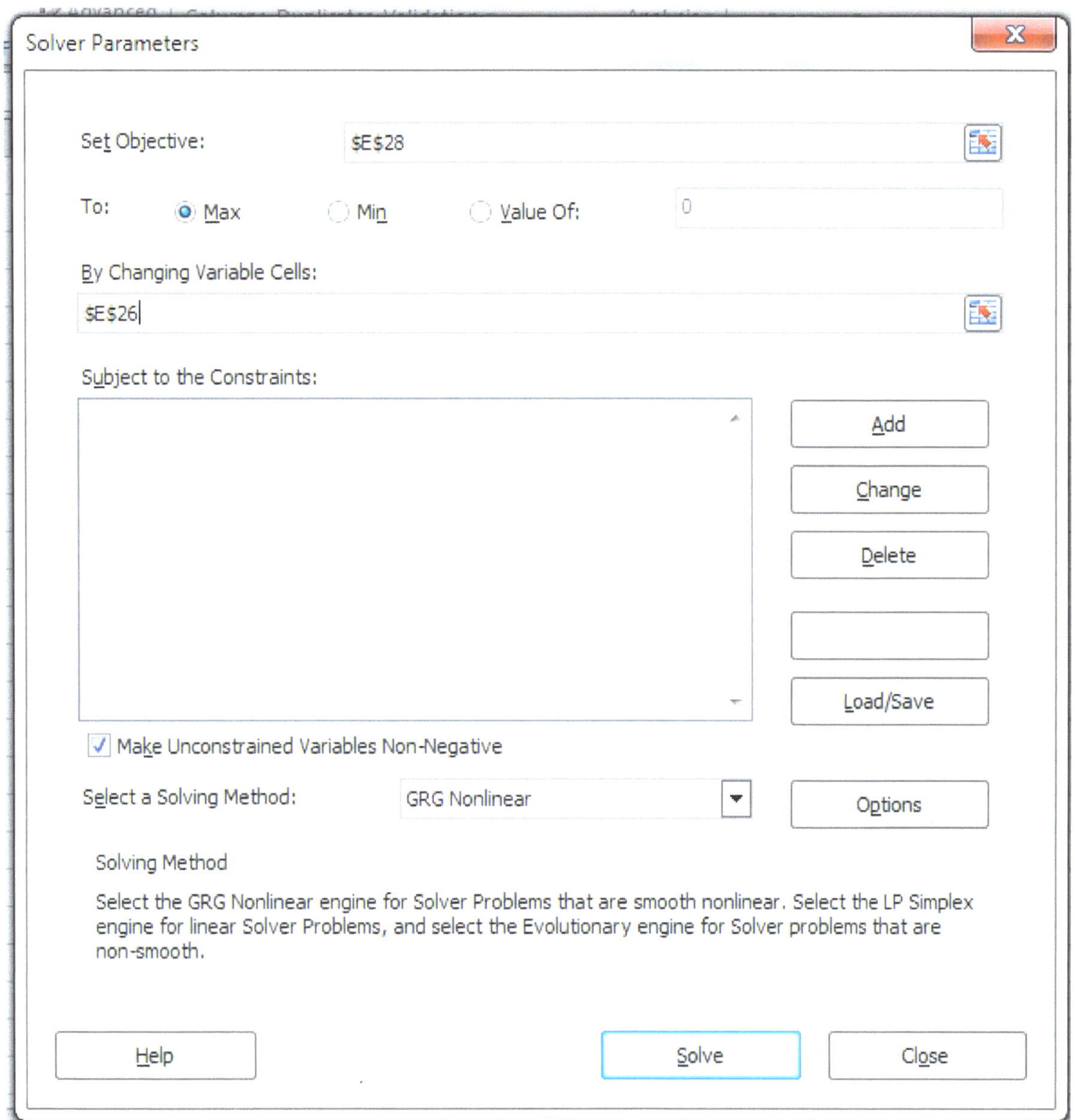

Figure 7.2.12

Click the Solve button.

Sell - Maximum Price

Month #:	44.940	= E26
Price, $:	28.467	= E28

Figure 7.2.13

So, based on the cubic least squares curve fit of the given stock data, the investor should have bought the stock at about week 13.5 at a price of about $20.88 and sold the stock at about week 45 at a price of about $28.47. Actual buy-sell prices and timing will vary dependent upon actual market activity around the indicated buy-sell timing periods.

PETROLEUM FRACTION BOILING POINT CURVES

WEB VIDEO ADDRESS: WWW.SONSHINECS.COM/VIDEOS/PETROLEUM_BOILING_POINTS_VIDEO.MP4

Laboratory analysis of a petroleum fraction has yielded the following boiling point distribution for the fraction:

Volume % Distilled	Temperature, °C
1	52
9	102
19	155
30	185
40	205
49	225
59	245
69	272
79	290
82	300
88	335
95	370
99	435

Table 7.3.1

It is desired to obtain an analytical equation representing these data so that several cuts can be made from the fraction at specific boiling point cut points, and so that the amounts of each cut can be determined. The desired cut points from the fraction are given in the following table:

Cut Range, °C
IBP - 71
71-177
177 - 204
204 - 274
274 - 316
316 - 343

343 - FBP

Table 7.3.2

We begin by entering the given laboratory analysis data into a Microsoft® Excel spreadsheet as follows.

Vol % Distilled	Temp °C
1	52
9	102
19	155
30	185
40	205
49	225
59	245
69	272
79	290
82	300
88	335
95	370
99	435

Figure 7.3.1

Since a typical boiling point curve for a petroleum fraction such as the one under consideration has an 'S' shape, we proceed to calculate the coefficients for a cubic least squares equation of the form $y = a_0 + a_1 x + a_2 x^2 + a_3 x^3$, where x is the volume percent distilled and y is the boiling point temperature in degrees C.

Using Excel to calculate the summations needed for the cubic least squares curve fit of the data, we have in Excel:

	x	x^2	x^3	x^4	x^5	x^6	y	xy	x^2y	x^3y
	1	1	1	1	1	1	52	52	52	52
	9	81	729	6561	59049	531441	102	918	8262	74358
	19	361	6859	130321	2476099	47045881	155	2945	55955	1063145
	30	900	27000	810000	24300000	729000000	185	5550	166500	4995000
	40	1600	64000	2560000	102400000	4096000000	205	8200	328000	13120000
	49	2401	117649	5764801	282475249	13841287201	225	11025	540225	26471025
	59	3481	205379	12117361	714924299	42180533641	245	14455	852845	50317855
	69	4761	328509	22667121	1564031349	1.07918E+11	272	18768	1294992	89354448
	79	6241	493039	38950081	3077056399	2.43087E+11	290	22910	1809890	142981310
	82	6724	551368	45212176	3707398432	3.04007E+11	300	24600	2017200	165410400
	88	7744	681472	59969536	5277319168	4.64404E+11	335	29480	2594240	228293120
	95	9025	857375	81450625	7737809375	7.35092E+11	370	35150	3339250	317228750
	99	9801	970299	96059601	9509900499	9.4148E+11	435	43065	4263435	422080065
Sums =	719	53121	4303679	365698185	32000149919	2.85688E+12	3171	217118	17270846	1461389528

Figure 7.3.2

Then, solving the cubic least squares relationships using the matrix manipulation functions MINVERSE() and MMULT() in Excel, we have

	13	719	53121	4303679		3171
A =	719	53121	4303679	365698185	B =	217118
	53121	4303679	365698185	32000149919		17270846
	4303679	365698185	32000149919	2.85688E+12		1461389528

	0.8431	-0.0610	0.0012	-6.59454E-06		40.998	$= a_0$
A^{-1} =	-0.0610	0.0070	-0.0002	9.79283E-07	X =	8.251	$= a_1$
	0.0012	-0.0002	0.0000	-2.50553E-08		-0.139	$= a_2$
	-6.6E-06	9.793E-07	-2.5055E-08	1.65577E-10		0.00095391	$= a_3$

Figure 7.3.3

So that the cubic least squares equation representing the laboratory boiling point data is

$$y = 40.998 + 8.251x - 0.139x^2 + 0.00095391x^3.$$

This equation represents the laboratory data in the following graph:

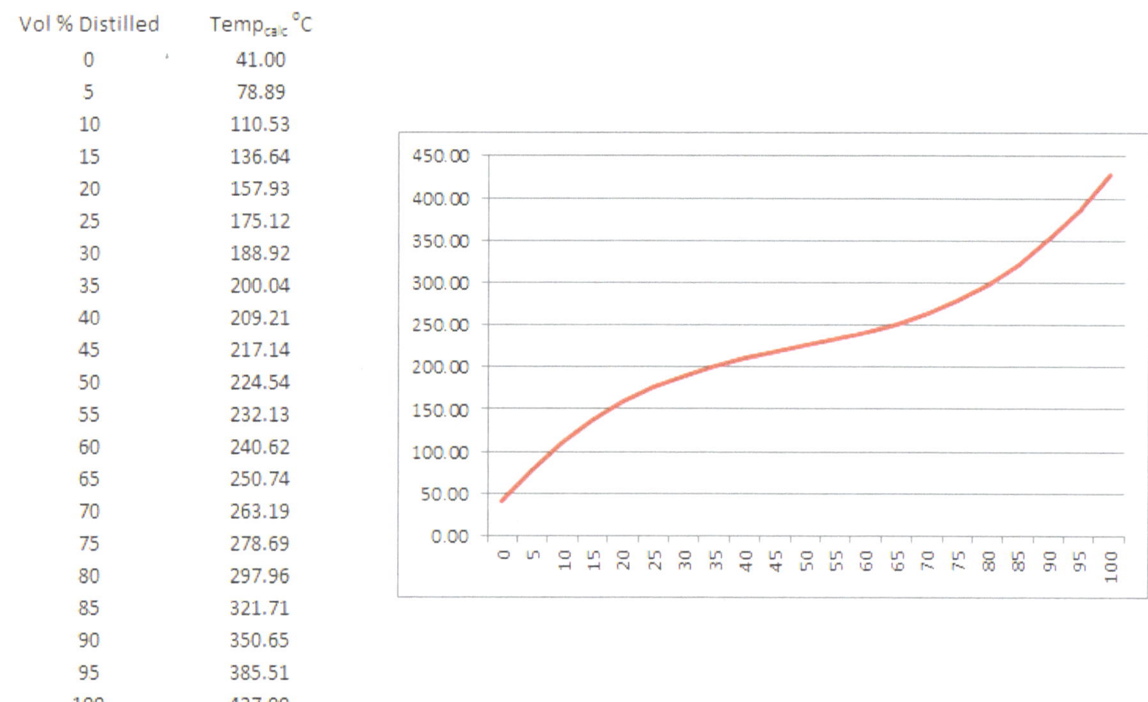

Vol % Distilled	$Temp_{calc}$ °C
0	41.00
5	78.89
10	110.53
15	136.64
20	157.93
25	175.12
30	188.92
35	200.04
40	209.21
45	217.14
50	224.54
55	232.13
60	240.62
65	250.74
70	263.19
75	278.69
80	297.96
85	321.71
90	350.65
95	385.51
100	427.00

Figure 7.3.4

Now, we use the cubic equation above to find the percent distilled (x) for each of the cut point temperatures (y) using Excel's Solver add-in. For example, to find the x value for the cut point $y = 177\,^{o}\mathrm{C}$, we enter the cubic equation into one Excel cell (cell K13, in this example) and some approximate x-value in another Excel cell (cell K12, in this example).

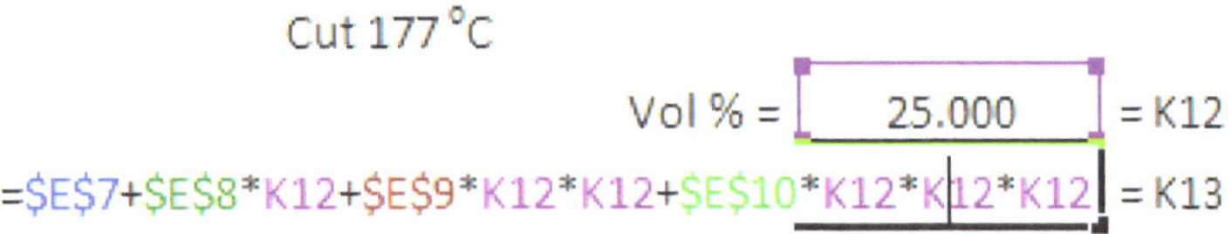

Figure 7.3.5

Using Solver, specify cell K13 as the Objective cell and cell K12 as the Variable cell as shown here.

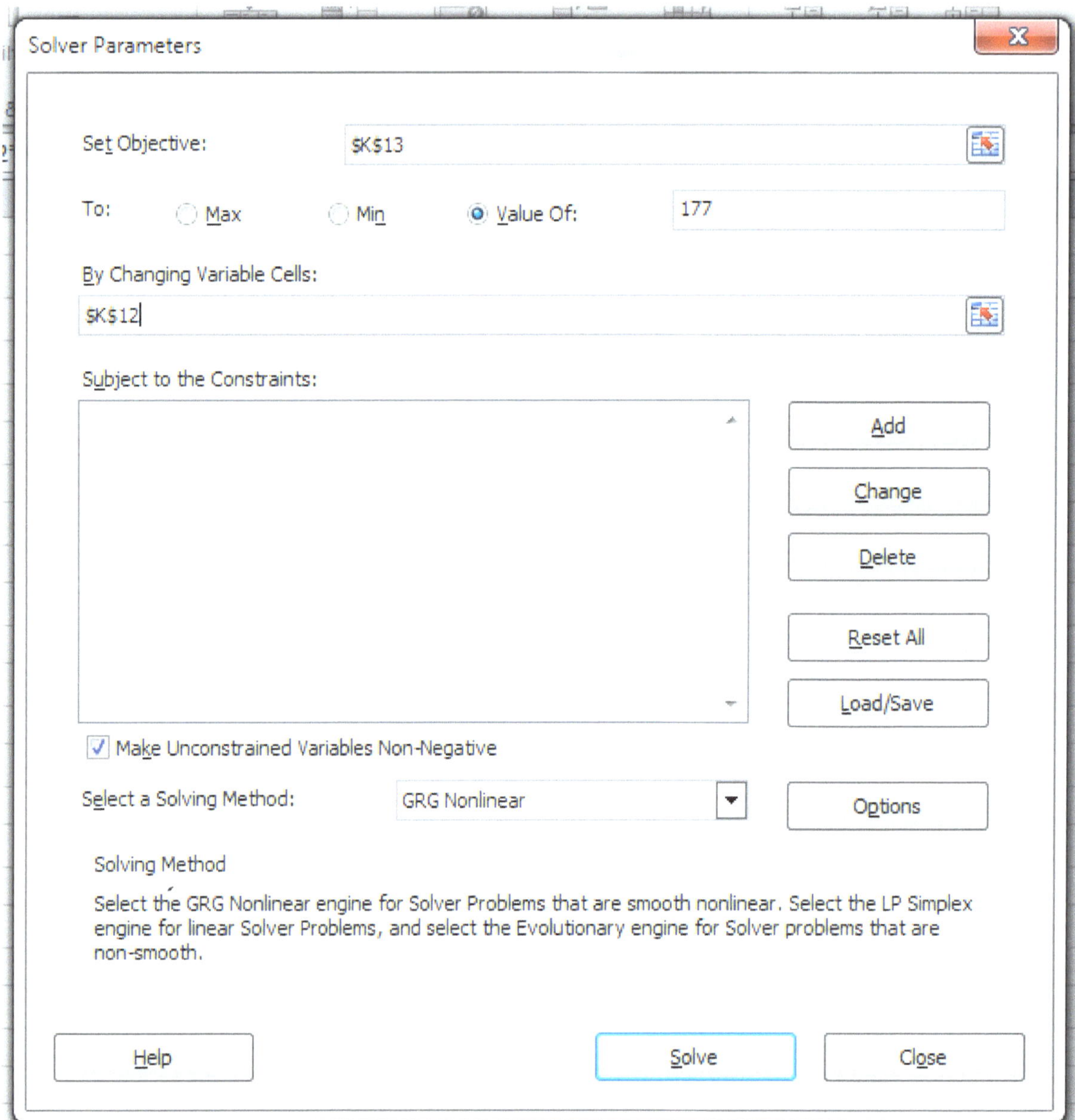

Figure 7.3.6

Click the Solve button.

Cut 177 °C

Vol % =	25.621	= K12
Temp, °C =	177.000	= K13

Figure 7.3.7

So that the volume percent (x) for a boiling point temperature (y) of 177 °C is 25.621%.

Similarly, to find the x value for the cut point $y = 316\,^{o}\mathrm{C}$, we enter the cubic equation into one Excel cell (cell K25, in this example) and some approximate x-value in another Excel cell (cell K24, in this example).

Cut 316 °C

Vol % = 80.000 = K24

=$E$7+$E$8*K24+$E$9*K24*K24+$E$10*K24*K24*K24 = K25

Figure 7.3.8

Using Solver, specifying cell K25 as the Objective cell and cell K24 as the Variable cell, we have

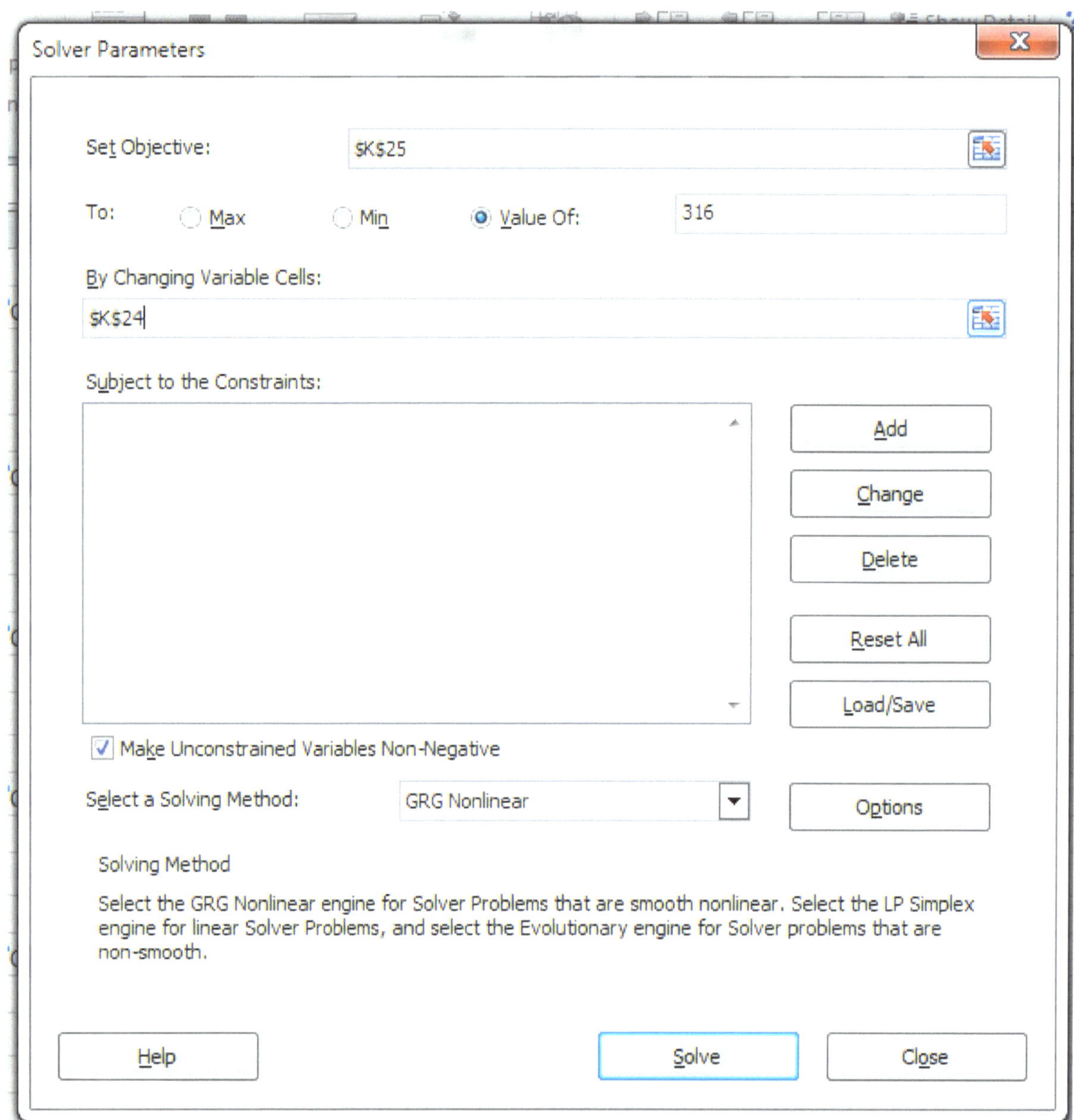

Figure 7.3.9

Click the Solve button.

Cut 316 °C

Vol % =	83.888	= K24
Temp, °C =	316.000	= K25

Figure 7.3.10

So that the volume percent (x) for a boiling point temperature (y) of 316 °C is 83.888%.

Repeating this process for each of the desired cut point temperatures and summarizing the cut point calculations, we have the volume percentages of each of the cuts from the original petroleum fraction.

Cut Point, °C	Pt. Vol %	Cut Vol %
IBP	0.00	0.000
71	3.88	3.88
177	25.62	21.74
204	36.52	10.90
274	73.29	36.77
316	83.89	10.60
343	88.77	4.88
FBP	100.00	11.23
	Sum =	100.00

Figure 7.3.11

CHURCH GROWTH

WEB VIDEO ADDRESS:

WWW.SONSHINECS.COM/VIDEOS/CHURCH_GROWTH_VIDEO.MP4

A church has recorded information about attendance at its Sunday worship service. The data show the average attendance at its worship service for each month for the past almost 4 years (47 months) as shown in the table below. The church would like to examine the recorded data to see if it gives some indication of whether or not attendance is growing so that it can plan for the growth in attendance either by adding a second Sunday worship service to its schedule or by planning an expansion of its worship facilities.

Month	Atten.	Month	Atten	Month	Atten	Month	Atten
1	437	13	495	25	452	37	599
2	495	14	598	26	597	38	454
3	474	15	477	27	403	39	439
4	461	16	489	28	459	40	596
5	473	17	481	29	437	41	489
6	445	18	449	30	406	42	460
7	457	19	427	31	415	43	419
8	421	20	369	32	523	44	428
9	412	21	487	33	637	45	391
10	456	22	442	34	412	46	427
11	511	23	467	35	423	47	495
12	500	24	670	36	412		

Table 7.4.1

To analyze the recorded data, we begin by using the method of least squares to fit the given data with a linear polynomial equation. We will use Microsoft® Excel for the calculations.

First, we enter the data into a table in Excel, a partial listing of which is shown below, and create a graph of the raw data as shown in the figure.

Month #	Attendance				
x	y	x	x^2	y	xy
1	437	1	1	437	437
2	495	2	4	495	990
3	474	3	9	474	1422
4	461	4	16	461	1844
5	473	5	25	473	2365
6	445	6	36	445	2670
7	457	7	49	457	3199
8	421	8	64	421	3368
9	412	9	81	412	3708
10	456	10	100	456	4560
11	511	11	121	511	5621
12	500	12	144	500	6000
13	495	13	169	495	6435
14	598	14	196	598	8372
15	477	15	225	477	7155
16	489	16	256	489	7824
17	481	17	289	481	8177
18	449	18	324	449	8082
19	427	19	361	427	8113

Figure 7.4.1

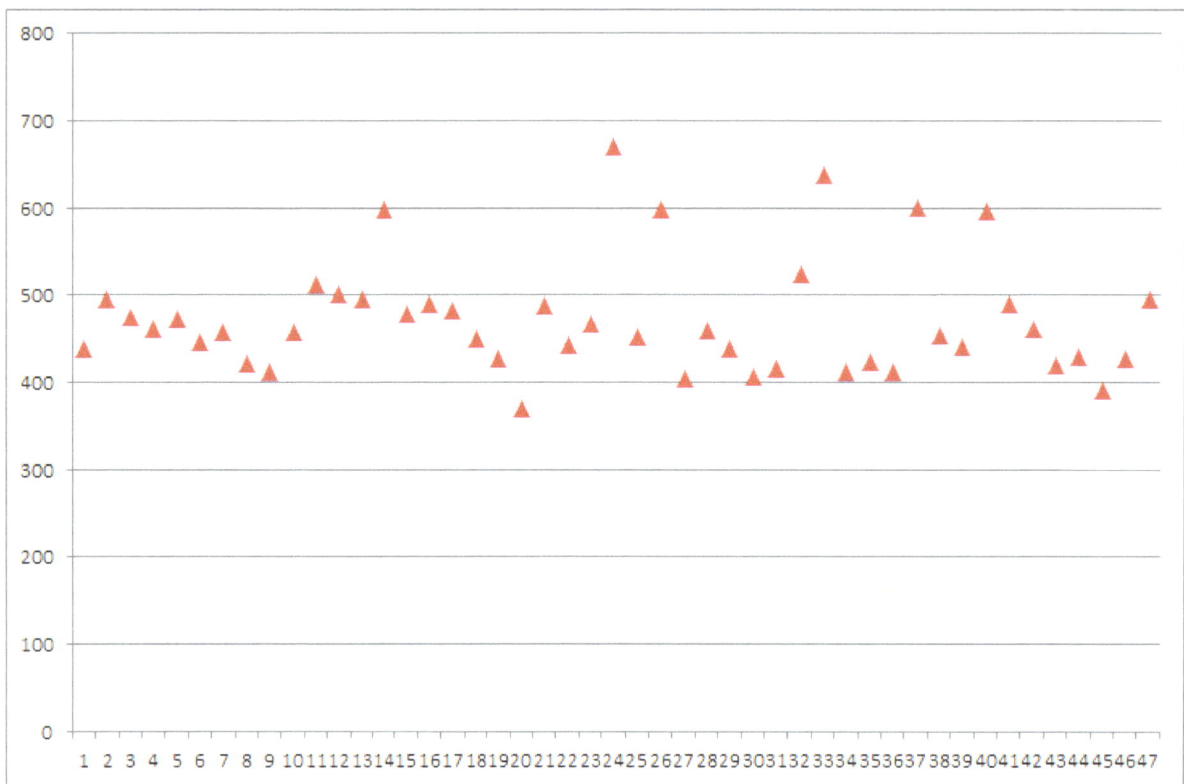

Figure 7.4.2

Performing a linear least squares calculation, we find the linear coefficients for the least squares straight-line equation as shown below.

$$A = \begin{bmatrix} N & \Sigma x \\ \Sigma x & \Sigma x^2 \end{bmatrix} \qquad B = \begin{bmatrix} \Sigma y \\ \Sigma xy \end{bmatrix}$$

$$A = \begin{bmatrix} 47 & 1128 \\ 1128 & 35720 \end{bmatrix} \qquad B = \begin{bmatrix} 22166 \\ 531265 \end{bmatrix}$$

$$A^{-1} = \begin{bmatrix} 0.0879 & -0.0028 \\ -0.0028 & 0.0001 \end{bmatrix} \qquad X = \begin{matrix} 473.6124 & = a_0 \\ -0.0831 & = a_1 \end{matrix}$$

Figure 7.4.3

So that the best straight line to represent the data based on the least squares criterion is $y = 473.61 - 0.0831x$.

Making a plot of the raw data versus the linear least squares line, we have the graph

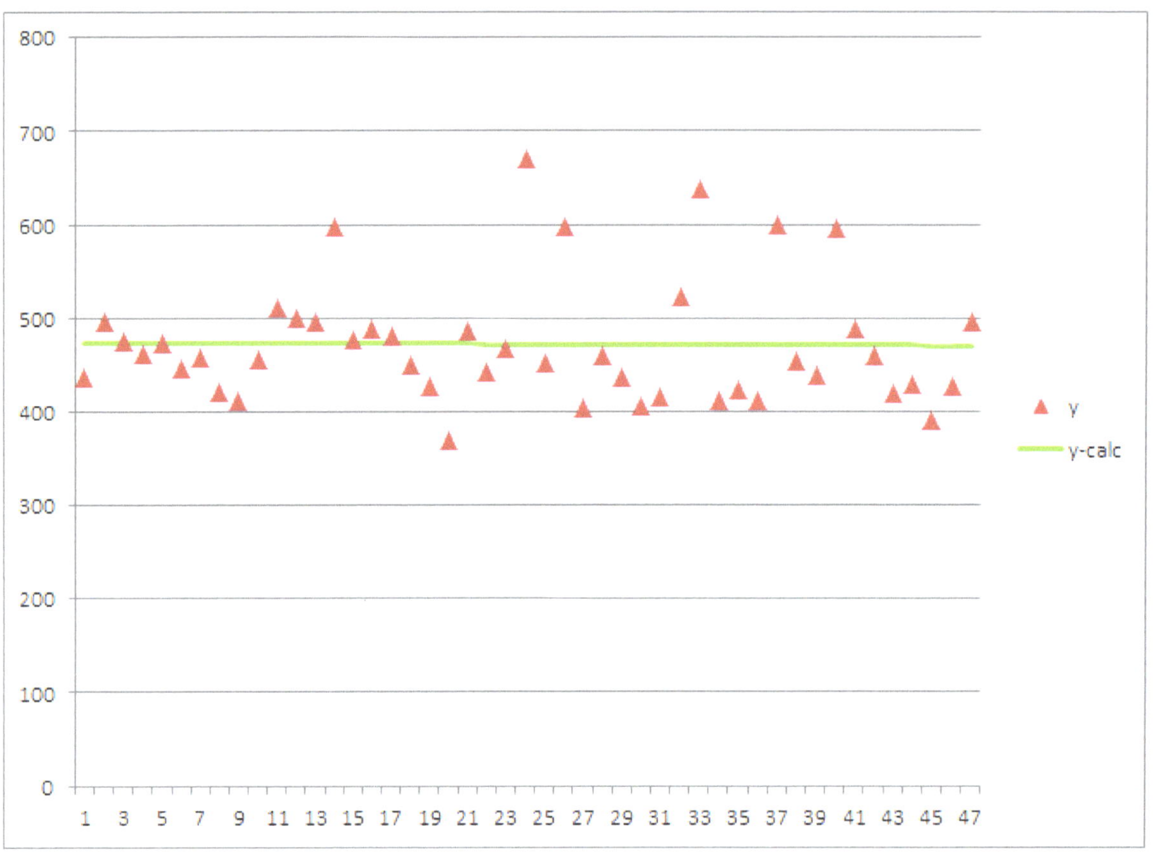

Figure 7.4.4

From this graph, we easily see that church attendance for the past four years has been essentially flat or constant with little indication of either growth or decline. In addition, we note that, on an overall average, attendance has been constant at about 473 worshippers per worship service. Noting further that the current permanent seating capacity of the worship facility is 590 (including choir loft seating), the current average attendance occupies almost exactly 80% of the available seating capacity of the worship facility. This information is consistent with Mylander's (4) findings that church growth either stops or slows dramatically when church attendance reaches approximately 75% of the capacity of the worship facility.

The conclusion that we reach from this analysis is that, since church attendance already exceeds the 75% level, the church needs to consider either adding a second Sunday worship service to its schedule or expanding the worship facility's seating capacity if the church is to grow beyond its current level of worship service attendance.

STEAKHOUSE RESTAURANT

WEB VIDEO ADDRESS: WWW.SONSHINECS.COM/VIDEOS/STEAKHOUSE_RESTAURANT_VIDEO.MP4

The Steakhouse Restaurant has recorded data of the number of meals it has served each month during the past twenty-four (24) months (see the table below). The manager would like to know the number of meals the restaurant can expect to serve during the next six (6) month period, assuming the trend of the past two years continues for the next six months.

NUMBER OF MEALS SERVED

Month	Number of Meals
January	9000
February	9500
March	10000
April	10700

May	10400
June	9900
July	10900
August	11200
September	10700
October	12100
November	12900
December	13300
January	12500
February	13500
March	14500
April	13900
May	15200
June	16600
July	19100
August	18100
September	19900
October	22500
November	23400
December	24200

Table 7.5.1

To make the projection or forecast for the next six months, we will use the method of least squares to fit the given data with linear, quadratic, and cubic polynomial equations. We will use Microsoft® Excel for the calculations.

To begin, we enter the data into a table in Excel and create a graph of the raw data as shown in the figure below.

Month #	Meals (1000's)
x	y
1	9.00
2	9.50
3	10.00
4	10.70
5	10.40
6	9.90
7	10.90
8	11.20
9	10.70
10	12.10
11	12.90
12	13.30
13	12.50
14	13.50
15	14.50
16	13.90
17	15.20
18	16.60
19	19.10
20	18.10
21	19.90
22	22.50
23	23.40
24	24.20

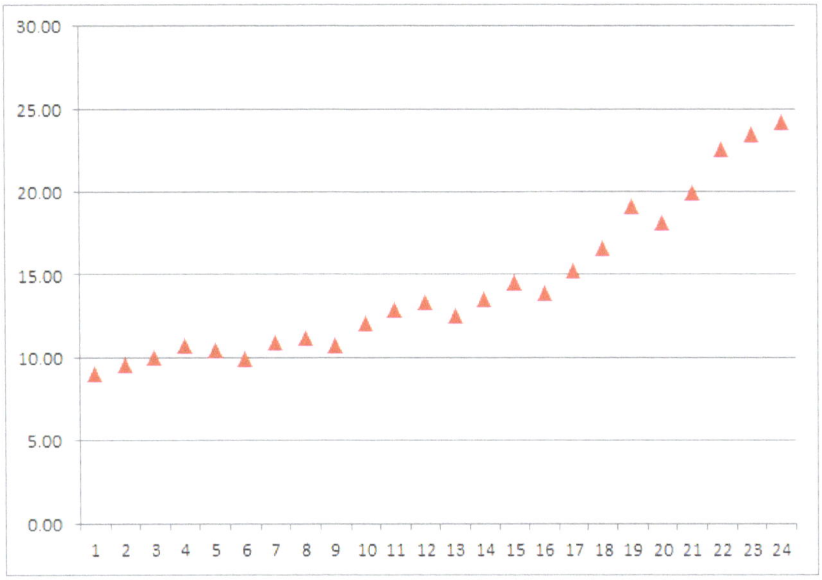

Figure 7.5.1

We calculate the coefficients of the linear least squares curve fit of the data of the form $y = a_0 + a_1 x$ in Excel as in Figure 7.5.2.

	x	x^2	y	xy
	1	1	9.00	9.00
	2	4	9.50	19.00
	3	9	10.00	30.00
	4	16	10.70	42.80
	5	25	10.40	52.00
	6	36	9.90	59.40
	7	49	10.90	76.30
	8	64	11.20	89.60
	9	81	10.70	96.30
	10	100	12.10	121.00
	11	121	12.90	141.90
	12	144	13.30	159.60
	13	169	12.50	162.50
	14	196	13.50	189.00
	15	225	14.50	217.50
	16	256	13.90	222.40
	17	289	15.20	258.40
	18	324	16.60	298.80
	19	361	19.10	362.90
	20	400	18.10	362.00
	21	441	19.90	417.90
	22	484	22.50	495.00
	23	529	23.40	538.20
	24	576	24.20	580.80
Sums =	300	4900	344	5002.3

A =	N	Σx	B =	Σy
	Σx	$Σx^2$		Σxy
A =	24	300	B =	344
	300	4900		5002.3
A^{-1} =	0.1775	-0.0109	X =	6.6996 $= a_0$
	-0.0109	0.0009		0.6107 $= a_1$

Figure 7.5.2

So the best straight line to represent the data based on the least squares criterion is $y = 6.6996 + 0.6107x$.

Similarly, we calculate the coefficients of the quadratic least squares curve fit of the data of the form $y = a_0 + a_1 x + a_2 x^2$ in Excel as in Figure 7.5.3.

	x	x^2	x^3	x^4	y	xy	x^2y
	1	1	1	1	9.00	9.00	9.00
	2	4	8	16	9.50	19.00	38.00
	3	9	27	81	10.00	30.00	90.00
	4	16	64	256	10.70	42.80	171.20
	5	25	125	625	10.40	52.00	260.00
	6	36	216	1296	9.90	59.40	356.40
	7	49	343	2401	10.90	76.30	534.10
	8	64	512	4096	11.20	89.60	716.80
	9	81	729	6561	10.70	96.30	866.70
	10	100	1,000	10000	12.10	121.00	1210.00
	11	121	1,331	14641	12.90	141.90	1560.90
	12	144	1,728	20736	13.30	159.60	1915.20
	13	169	2,197	28561	12.50	162.50	2112.50
	14	196	2,744	38416	13.50	189.00	2646.00
	15	225	3,375	50625	14.50	217.50	3262.50
	16	256	4,096	65536	13.90	222.40	3558.40
	17	289	4,913	83521	15.20	258.40	4392.80
	18	324	5,832	104976	16.60	298.80	5378.40
	19	361	6,859	130321	19.10	362.90	6895.10
	20	400	8,000	160000	18.10	362.00	7240.00
	21	441	9,261	194481	19.90	417.90	8775.90
	22	484	10,648	234256	22.50	495.00	10890.00
	23	529	12,167	279841	23.40	538.20	12378.60
	24	576	13,824	331776	24.20	580.80	13939.20
Sums =	300	4900	90000	1763020	344	5002.30	89197.70

	24	300	4900		344	
A =	300	4900	90000	B =	5002.3	
	4900	90000	1763020		89197.7	
	0.4452	-0.0726	0.0025		10.175	$= a_0$
A^{-1} =	-0.0726	0.0151	-0.0006	X =	-0.191	$= a_1$
	0.0025	-0.0006	0.0000		0.032	$= a_2$

Figure 7.5.3

From which we get the best quadratic equation to represent the data based on the least squares criterion is $y = 10.175 - 0.191x + 0.032x^2$.

We use Excel to calculate the coefficients of the cubic least squares curve fit of the data of the form $y = a_0 + a_1 x + a_2 x^2 + a_3 x^3$ as in Figure 7.5.4 below.

	x	x^2	x^3	x^4	x^5	x^6	y	xy	x^2y	x^3y
	1	1	1	1	1	1	9.00	9.00	9.00	9.00
	2	4	8	16	32	64	9.50	19.00	38.00	76.00
	3	9	27	81	243	729	10.00	30.00	90.00	270.00
	4	16	64	256	1,024	4096	10.70	42.80	171.20	684.80
	5	25	125	625	3,125	15625	10.40	52.00	260.00	1300.00
	6	36	216	1296	7,776	46656	9.90	59.40	356.40	2138.40
	7	49	343	2401	16,807	117649	10.90	76.30	534.10	3738.70
	8	64	512	4096	32,768	262144	11.20	89.60	716.80	5734.40
	9	81	729	6561	59,049	531441	10.70	96.30	866.70	7800.30
	10	100	1,000	10000	100,000	1000000	12.10	121.00	1210.00	12100.00
	11	121	1,331	14641	161,051	1771561	12.90	141.90	1560.90	17169.90
	12	144	1,728	20736	248,832	2985984	13.30	159.60	1915.20	22982.40
	13	169	2,197	28561	371,293	4826809	12.50	162.50	2112.50	27462.50
	14	196	2,744	38416	537,824	7529536	13.50	189.00	2646.00	37044.00
	15	225	3,375	50625	759,375	11390625	14.50	217.50	3262.50	48937.50
	16	256	4,096	65536	1,048,576	16777216	13.90	222.40	3558.40	56934.40
	17	289	4,913	83521	1,419,857	24137569	15.20	258.40	4392.80	74677.60
	18	324	5,832	104976	1,889,568	34012224	16.60	298.80	5378.40	96811.20
	19	361	6,859	130321	2,476,099	47045881	19.10	362.90	6895.10	131006.90
	20	400	8,000	160000	3,200,000	64000000	18.10	362.00	7240.00	144800.00
	21	441	9,261	194481	4,084,101	85766121	19.90	417.90	8775.90	184293.90
	22	484	10,648	234256	5,153,632	1.13E+08	22.50	495.00	10890.00	239580.00
	23	529	12,167	279841	6,436,343	1.48E+08	23.40	538.20	12378.60	284707.80
	24	576	13,824	331776	7,962,624	1.91E+08	24.20	580.80	13939.20	334540.80
Sums =	300	4900	90000	1763020	35970000	7.55E+08	344	5002.30	89197.70	1734800.50

	24	300	4900	90000		344	
A =	300	4900	90000	1763020	B =	5002.3	
	4900	90000	1763020	35970000		89197.7	
	90000	1763020	35970000	754740700		1734801	
	0.9269	-0.2827	0.0231	-0.000549		8.854	$= a_0$
A^{-1} =	-0.2827	0.1067	-0.0095	0.0002394	X =	0.385	$= a_1$
	0.0231	-0.0095	0.0009	-2.35E-05		-0.024	$= a_2$
	-0.00055	0.000239	-2.34602E-05	6.256E-07		0.001506	$= a_3$

Figure 7.5.4

From which we get that the best cubic equation to represent the data based on the least squares criterion is

$$y = 8.8538 + 0.3849x - 0.02439x^2 + 0.001506x^3.$$

The representation of the given data by the three curve fits (linear, quadratic, and cubic) is shown on the following graph (Figure 7.5.5):

Month #		Meals (1000's)		
x	y	y_{linear}	$y_{quadratic}$	y_{cubic}
1	9.00	7.310	10.016	9.216
2	9.50	7.921	9.921	9.538
3	10.00	8.532	9.890	9.830
4	10.70	9.142	9.923	10.100
5	10.40	9.753	10.020	10.357
6	9.90	10.364	10.182	10.611
7	10.90	10.975	10.408	10.870
8	11.20	11.585	10.698	11.143
9	10.70	12.196	11.052	11.441
10	12.10	12.807	11.470	11.770
11	12.90	13.417	11.952	12.142
12	13.30	14.028	12.499	12.563
13	12.50	14.639	13.109	13.045
14	13.50	15.249	13.784	13.595
15	14.50	15.860	14.523	14.223
16	13.90	16.471	15.327	14.938
17	15.20	17.081	16.194	15.748
18	16.60	17.692	17.125	16.663
19	19.10	18.303	18.121	17.692
20	18.10	18.914	19.181	18.844
21	19.90	19.524	20.305	20.128
22	22.50	20.135	21.493	21.553
23	23.40	20.746	22.745	23.128
24	24.20	21.356	24.062	24.862

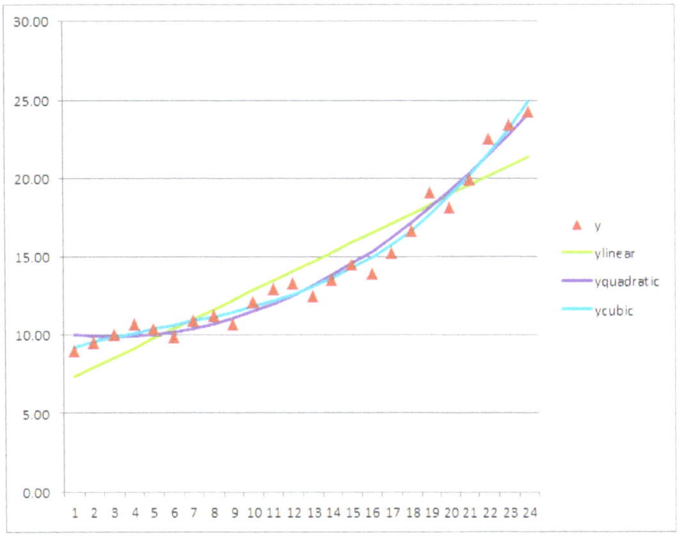

Figure 7.5.5

Now, we use the three least squares equations to project or forecast six months into the future. Another way of stating this is to say that we want to use the three equations, linear, quadratic, and cubic, to calculate the number of meals to be served per month for months 25 through 30; that is, we want to extrapolate the data for the six months following the 24 months for which we have meal service data. This forecast calculation, performed using Excel with the mathematical models derived above, is shown below.

Curve Fits

	Linear	Quadratic	Cubic
a_0 =	6.699637681	10.17509881	8.853754941
a_1 =	0.610695652	-0.19133384	0.384937727
a_2 =		0.03208118	-0.024386507
a_3 =			0.001505805

Month	Month #		Meals (1000's)	
	x	y_{linear}	$y_{quadratic}$	y_{cubic}
Jan	25	21.967	25.442	26.764
Feb	26	22.578	26.887	28.843
Mar	27	23.188	28.396	31.108
Apr	28	23.799	29.969	33.568
May	29	24.410	31.607	36.233
Jun	30	25.021	33.308	39.111

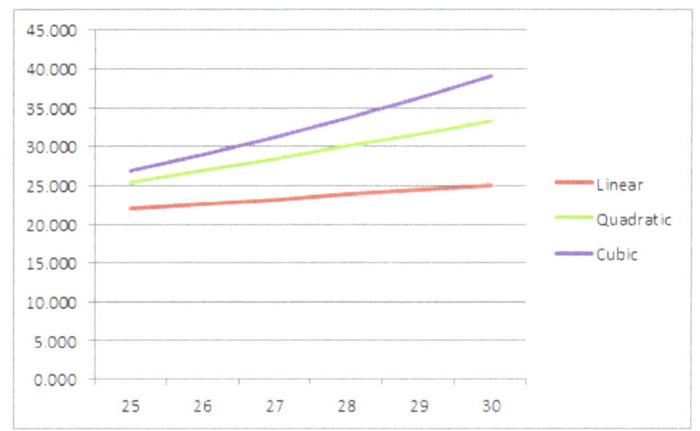

Figure 7.5.6

So, based on a least squares criterion for representing the given data and using those representations to project or forecast for the next six months, the restaurant can expect to serve between approximately 25,000 and 39,000 meals per month by the end of the sixth month.

APPENDIX

APPENDIX 1: SYMBOLS, NOMENCLATURE, MATHEMATICAL NOTATION, MICROSOFT® EXCEL FUNCTIONS

* - indicates multiplication of elements in a Microsoft® Excel formula

- - indicates subtraction of elements in a Microsoft® Excel formula

/ - indicates division of elements in a Microsoft® Excel formula

+ - indicates addition of elements in a Microsoft® Excel formula

<Ctrl><Shift><Enter> - hold down the <Ctrl> and <Shift> keys (easily done with the left hand in the lower left of the keyboard) and press the <Enter> key (with the right hand).

$\mathbf{A}$ - a matrix

$\mathbf{A}^{-1}$ - inverse of the $\mathbf{A}$ matrix

$\mathbf{B}$ - the right-hand-side (constant) vector in a matrix expression

$\mathbf{X}$ - the unknown vector in a matrix expression

COUNT() – Microsoft® Excel function for counting the number of elements in a specified range of Excel cells

MINVERSE() – Microsoft® Excel function for calculating the inverse of a matrix

MMULT() – Microsoft® Excel function for multiplying two matrices

N - the number of data points to be fitted with the least squares polynomial

SUM() – Microsoft® Excel function for adding the values contained in a specified range of Excel cells

∂ - partial derivative operator

i, k, n, N - subscripts

x - independent variable, normally plotted along the horizontal axis in a rectangular coordinate system

y - dependent variable, normally plotted along the vertical axis in a rectangular coordinate system

$\sum$ - summation: indicates the summation of the specified items

APPENDIX 2: DEVELOPMENT OF AND THEORETICAL BASIS FOR THE METHOD OF LEAST SQUARES

The Method of Least Squares is a mathematical procedure for representing a set or table of data by an analytical mathematical equation, and is of interest for interpolating and/or extrapolating the set of data using a smoothed representation of the original data. The method is frequently presented as a way of representing the set of data with a linear or straight-line analytical equation, but, as we will show in the following discussion, it can also be applied to represent the data with higher-order polynomial equations such as the quadratic and cubic forms.

To begin our discussion, consider a set of two-dimensional data represented graphically below. Here, x is the independent variable normally plotted along the horizontal axis and y is the dependent variable normally plotted along the vertical axis in a rectangular coordinate system. The individual data points are represented by the general coordinates $\left(x_k, y_k\right)$.

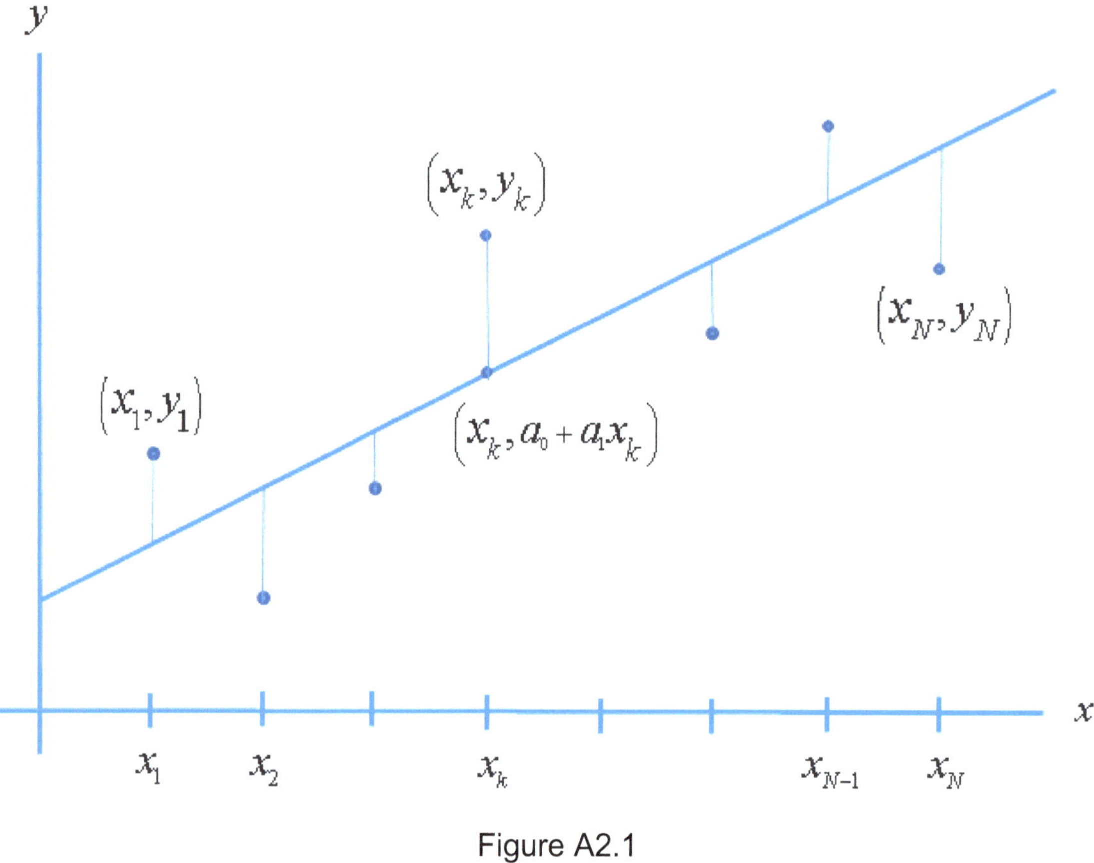

Figure A2.1

Suppose it is desired to fit these data with a straight line equation of the form $y = a_0 + a_1 x$ where a_1 is the slope of the straight line and a_0 is the y-intercept of the line.

There are obviously many straight lines that could be drawn through the given data. What we want is the “best” straight line through the data where “best” is defined by some quantitative mathematical criterion. The criterion we will use is called a least squares criterion, and is described as that straight line such that the sum of the squares of the differences is a minimum.

Now, what do we mean by saying that "the sum of the squares of the differences is a minimum"?

To answer this question, we note first of all that, in the graph above, for each value of x (that is, for each x_k), there are two values for y corresponding to that particular x value: the first is the given value of y (y_k) corresponding to the original given data, while the second is the value of y calculated from the straight line equation, where the calculated value of y is given by the equation $y = a_0 + a_1 x_k$. The difference between these two values of y for any $x = x_k$ is $a_0 + a_1 x_k - y_k$. Note that x_k and y_k are known values since their values come directly from the coordinates of the given original data.

The least squares criterion says that the best straight-line equation is that straight line equation such that, if we take the difference between the two values of y corresponding to every x_k, square that difference, and then sum the squares of the differences, the sum of all these differences is a minimum or as small as possible. This, then will define the 'best' straight line that can be drawn through the given data, and therefore will be the 'best' straight line that we can use to represent the given data based on the least squares criterion.

To express the criterion described above in quantitative mathematical terms and in such a way as to be able to minimize the sum of the squares of the differences, we can first represent each of the differences as follows:

$$\left(a_0 + a_1 x_1 - y_1\right)$$

$$\left(a_0 + a_1 x_2 - y_2\right)$$

....

$$\left(a_0 + a_1 x_n - y_n\right)$$

Next, we let E be the sum of the squares of the differences, so that we have

$$E = \left(a_0 + a_1 x_1 - y_1\right)^2 + \left(a_0 + a_1 x_2 - y_2\right)^2 + \left(a_0 + a_1 x_3 - y_3\right)^2 + \ldots\ldots\ldots + \left(a_0 + a_1 x_N - y_N\right)^2$$

where N is the number of original data points. In summation notation form, the sum of the squares of the differences is

$$E = \sum_{i = 1}^{i = N} \left(a_0 + a_1 x_i - y_i\right)^2$$

Note that the x_i's and y_i's are all known values from the original set of data. The unknowns in the expression for E are the a_0 and a_1 values; that is, the unknowns are the coefficients of the straight line (first-order polynomial) equation $y = a_0 + a_1 x_k$.

In order to minimize the function E, we need to set the partial derivative of E with respect to each of the unknowns equal to zero and solve the resulting equations simultaneously. That is, we must solve the equations

$$\frac{\delta E}{\delta a_0} = 0$$

$$\frac{\delta E}{\delta a_1} = 0$$

simultaneously. The partial derivatives are:

$$\frac{\partial E}{\partial a_0} = 2\left(a_0 + a_1 x_1 - y_1\right) + 2\left(a_0 + a_1 x_2 - y_2\right)$$

$$+ 2\left(a_0 + a_1 x_3 - y_3\right) + \ldots\ldots\ldots$$

$$\frac{\partial E}{\partial a_0} = \sum_{i=1}^{N} 2a_0 + \sum_{1}^{N} 2a_1 x_i - \sum_{1}^{N} 2y_i$$

and

$$\frac{\partial E}{\partial a_1} = 2x_1\left(a_0 + a_1 x_1 - y_1\right) + 2x_2\left(a_0 + a_1 x_2 - y_2\right)$$

$$+ 2x_3\left(a_0 + a_1 x_3 - y_3\right) + \ldots\ldots\ldots$$

$$\frac{\partial E}{\partial a_1} = \sum_{i=1}^{N} 2a_0 x_i + \sum_{1}^{N} 2a_1 x_i^2 - \sum_{1}^{N} 2x_i y_i$$

If we now set the partial derivatives above equal to zero, we have

$$\sum_{i=1}^{N} 2a_0 + \sum_{1}^{N} 2a_1 x_i = \sum_{1}^{N} 2y_i$$

and

$$\sum_{i=1}^{N} 2a_0 x_i + \sum_{1}^{N} 2a_1 x_i^2 = \sum_{1}^{N} 2x_i y_i$$

or

$$a_0\sum_{i=1}^{N}1 + a_1\sum_{1}^{N}x_i = \sum_{1}^{N}y_i$$

and

$$a_0\sum_{i=1}^{N}x_i + a_1\sum_{1}^{N}x_i^2 = \sum_{1}^{N}x_iy_i$$

Thus, the result of taking the partial derivatives and setting them equal to zero is given by the following equations:

$$a_0 N \qquad + a_1\sum_{i=1}^{i=N}(x_i) = \sum_{i=1}^{i=N}(y_i)$$

$$a_0\sum_{i=1}^{i=N}(x_i) + a_1\sum_{i=1}^{i=N}(x_i^2) = \sum_{i=1}^{i=N}(x_i\,y_i)$$

where N is the number of pairs of x, y values in the original set of data, $\sum_{i=1}^{i=N}(x_i)$ is the sum of all of the given x-values in the original data set, $\sum_{i=1}^{i=N}(x_i^2)$ is the sum of all the given x-values squared, $\sum_{i=1}^{i=N} (y_i)$ is the sum of all the given y-values, and $\sum_{i=1}^{i=N} (x_i \, y_i)$ is the sum of all of the given x, y-value products.

These may look like rather complex equations to solve. However, if we note that all of the summations are simply numbers, numerical values derived from the data in the original set of data, we can also note that these are two linear equations in two unknowns, a_0 and a_1, that can be solved either by typical analytical methods (substitution, for example) or by using matrix manipulation techniques.

The development described above can easily be extended to fitting the data with the best quadratic (second-order polynomial) or the best cubic (third-order polynomial) equation to represent the data based on the least squares criterion. We will illustrate the development of the quadratic fit and leave the corresponding cubic development as an exercise for the reader.

To express the least squares criterion for a quadratic fit of the data, where the equation to be used to represent the data is a quadratic equation of the form $y = a_0 + a_1 x + a_2 x^2$, we first represent each of the differences in the y-values for each individual x-value as follows:

$$\left(a_0 + a_1 x_1 + a_2 x_1^2 - y_1\right)$$

$$\left(a_0 + a_1 x_2 + a_2 x_2^2 - y_2\right)$$

....

$$\left(a_0 + a_1 x_n + a_2 x_n^2 - y_n\right)$$

Next, we let E be the sum of the squares of the differences, so that we have

$$\begin{aligned} E = & \left(a_0 + a_1 x_1 + a_2 x_1^2 - y_1\right)^2 \\ & + \left(a_0 + a_1 x_2 + a_2 x_2^2 - y_2\right)^2 \\ & + \left(a_0 + a_1 x_3 + a_2 x_3^2 - y_3\right)^2 + \ldots\ldots\ldots \\ & + \left(a_0 + a_1 x_N + a_2 x_N^2 - y_N\right)^2 \end{aligned}$$

where N is the number of original data points. In summation notation form,

$$E = \sum_{i=1}^{i=N} \left(a_0 + a_1 x_i + a_2 x_i^2 - y_i\right)^2$$

Once again, note that the x_i's and y_i's are all known values from the original set of data. The unknowns in the expression for E are the a_0, a_1, and a_2 values; that is, the coefficients of the quadratic (second-order polynomial) equation $y = a_0 + a_1 x_k + a_2 x_k^2$.

In order to minimize the function E, we need to set the partial derivative of E with respect to each of the unknowns equal to zero and solve the resulting equations simultaneously. That is, for the quadratic least squares fit, we must solve the equations

$$\frac{\delta E}{\delta a_0} = 0$$

$$\frac{\delta E}{\delta a_1} = 0$$

$$\frac{\delta E}{\delta a_2} = 0$$

simultaneously. The partial derivatives are

$$\frac{\partial E}{\partial a_0} = 2\left(a_0 + a_1 x_1 + a_2 x_1^2 - y_1\right) + 2\left(a_0 + a_1 x_2 + a_2 x_2^2 - y_2\right)$$

$$+ 2\left(a_0 + a_1 x_3 + a_2 x_3^2 - y_3\right) + \ldots\ldots$$

$$\frac{\partial E}{\partial a_0} = \sum_{i=1}^{N} 2a_0 + \sum_{1}^{N} 2a_1 x_i + \sum_{1}^{N} 2a_2 x_i^2 - \sum_{1}^{N} 2y_i$$

$$\frac{\partial E}{\partial a_1} = 2x_1\left(a_0 + a_1 x_1 + a_2 x_1^2 - y_1\right)$$

$$+ 2x_2\left(a_0 + a_1 x_2 + a_2 x_2^2 - y_2\right)$$

$$+ 2x_3\left(a_0 + a_1 x_3 + a_2 x_3^2 - y_3\right) + \ldots\ldots$$

$$\frac{\partial E}{\partial a_1} = \sum_{i=1}^{N} 2a_0 x_i + \sum_{1}^{N} 2a_1 x_i^2 + \sum_{1}^{N} 2a_2 x_i^3 - \sum_{1}^{N} 2x_i y_i$$

and

$$\frac{\partial E}{\partial a_2} = 2x_1^2\left(a_0 + a_1 x_1 + a_2 x_1^2 - y_1\right)$$

$$+ 2x_2^2\left(a_0 + a_1 x_2 + a_2 x_2^2 - y_2\right)$$

$$+ 2x_3^2\left(a_0 + a_1 x_3 + a_2 x_3^2 - y_3\right) + \ldots\ldots$$

$$\frac{\partial E}{\partial a_2} = \sum_{i=1}^{N} 2a_0 x_i^2 + \sum_{1}^{N} 2a_1 x_i^3 + \sum_{1}^{N} 2a_2 x_i^4 - \sum_{1}^{N} 2x_i^2 y_i$$

and when setting the partial derivatives equal to zero, we have

$$\sum_{i=1}^{N} 2a_0 \quad + \sum_{1}^{N} 2a_1 x_i + \sum_{1}^{N} 2a_2 x_i^2 = \sum_{1}^{N} 2y_i$$

$$\sum_{i=1}^{N} 2a_0 x_i + \sum_{1}^{N} 2a_1 x_i^2 + \sum_{1}^{N} 2a_2 x_i^3 = \sum_{1}^{N} 2x_i y_i$$

and

$$\sum_{i=1}^{N} 2a_0 x_i^2 + \sum_{1}^{N} 2a_1 x_i^3 + \sum_{1}^{N} 2a_2 x_i^4 = \sum_{1}^{N} 2x_i^2 y_i$$

or

$$a_0 \sum_{i=1}^{N} 1 \quad + a_1 \sum_{1}^{N} x_i + a_2 \sum_{1}^{N} x_i^2 = \sum_{1}^{N} y_i$$

$$a_0 \sum_{i=1}^{N} x_i + a_1 \sum_{1}^{N} x_i^2 + a_2 \sum_{1}^{N} x_i^3 = \sum_{1}^{N} x_i y_i$$

and

$$a_0 \sum_{i=1}^{N} x_i^2 + a_1 \sum_{1}^{N} x_i^3 + a_2 \sum_{1}^{N} x_i^4 = \sum_{1}^{N} x_i^2 y_i$$

Thus, the result of taking the partial derivatives and setting them equal to zero can be stated by the following equations.

$$a_0 N + a_1 \sum_{i=1}^{i=N} (x_i) + a_2 \sum_{i=1}^{i=N} (x_i^2) = \sum_{i=1}^{i=N} (y_i)$$

$$a_0 \sum_{i=1}^{i=N} (x_i) + a_1 \sum_{i=1}^{i=N} (x_i^2) + a_2 \sum_{i=1}^{i=N} (x_i^3) = \sum_{i=1}^{i=N} (x_i \, y_i)$$

$$a_0 \sum_{i=1}^{i=N} (x_i^2) + a_1 \sum_{i=1}^{i=N} (x_i^3) + a_2 \sum_{i=1}^{i=N} (x_i^4) = \sum_{i=1}^{i=N} (x_i^2 \, y_i)$$

where N is the number of pairs of x, y values in the original set of data, $\sum_{i=1}^{i=N} (x_i)$ is the sum of all of the given x-values in the original data set, $\sum_{i=1}^{i=N} (x_i^2)$ is the sum of all the given x-values squared, $\sum_{i=1}^{i=N} (x_i^3)$ is the sum of all the given x-values cubed, $\sum_{i=1}^{i=N} (x_i^4)$ is the sum of all the given x-values raised to the fourth power, $\sum_{i=1}^{i=N} (y_i)$ is the sum of all the given y-values, $\sum_{i=1}^{i=N} (x_i \, y_i)$ is the sum of all of the given x, y-value products, and $\sum_{i=1}^{i=N} \left(x_i^2 \, y_i \right)$ is the sum of all of the given x^2, y -value products.

Once again, these may look like rather complex equations to solve. However, if we note that all of the summations are simply numbers, numerical values derived from the data in the original set of data, we can also note that these are three linear equations in three unknowns, a_0, a_1, and a_2. These three equations in three unknowns can be solved by typical analytical methods of substitution, but the process is prone to errors in implementing the substitutions and is somewhat tedious to perform. On the other hand, the equations can be solved easily by using matrix manipulation techniques such as those found in Microsoft® Excel and other computer applications.

In simplified notation, the equations that need to be solved are

$$a_0 N \quad + a_1 \sum (x) \ + a_2 \sum \left(x^2\right) = \sum (y)$$

$$a_0 \sum (x) \ + a_1 \sum \left(x^2\right) + a_2 \sum \left(x^3\right) = \sum (xy)$$

$$a_0 \sum \left(x^2\right) + a_1 \sum \left(x^3\right) + a_2 \sum \left(x^4\right) = \sum \left(x^2 y\right)$$

where the quadratic equation used to fit the data is of the form

$$y = a_0 + a_1 x + a_2 x^2$$

Similarly, if the data are to be represented by a cubic least squares curve fit, the coefficients of the cubic equation (third-order polynomial) are obtained by solving the following equations simultaneously:

$$a_0 N \quad + a_1 \Sigma(x) \; + a_2 \Sigma\left(x^2\right) + a_3 \Sigma\left(x^3\right) = \Sigma(y)$$
$$a_0 \Sigma(x) \; + a_1 \Sigma\left(x^2\right) + a_2 \Sigma\left(x^3\right) + a_3 \Sigma\left(x^4\right) = \Sigma(xy)$$
$$a_0 \Sigma\left(x^2\right) + a_1 \Sigma\left(x^3\right) + a_2 \Sigma\left(x^4\right) + a_3 \Sigma\left(x^5\right) = \Sigma\left(x^2 y\right)$$
$$a_0 \Sigma\left(x^3\right) + a_1 \Sigma\left(x^4\right) + a_2 \Sigma\left(x^5\right) + a_3 \Sigma\left(x^6\right) = \Sigma\left(x^3 y\right)$$

where the cubic equation used to fit the original data is of the form

$$y = a_0 + a_1 x + a_2 x^2 + a_3 x^3$$

Indeed, these relationships can easily be extended to any higher order polynomial least squares data fit simply by noting the patterns represented in the equations above.

APPENDIX 3: SUMMARY OF LEAST SQUARES RELATIONSHIPS

LINEAR LEAST SQUARES:

$$y = a_0 + a_1 x$$

$$a_0 N \quad + a_1 \sum(x) \; = \sum(y)$$
$$a_0 \sum(x) + a_1 \sum(x^2) = \sum(xy)$$

Matrix Representation:

$$\mathbf{AX} = \mathbf{B}, \qquad \mathbf{X} = \mathbf{A}^{-1}\mathbf{B}$$

$$\mathbf{A} = \begin{bmatrix} N & \sum(x) \\ \sum(x) & \sum(x^2) \end{bmatrix}, \quad \mathbf{X} = \begin{bmatrix} a_0 \\ a_1 \end{bmatrix}, \quad \mathbf{B} = \begin{bmatrix} \sum(y) \\ \sum(xy) \end{bmatrix}$$

Analytically:

$$a_0 = \frac{\sum(x^2)\sum(y) - \sum(x)\sum(xy)}{N\sum(x^2) - \sum(x)\sum(x)}$$

$$a_1 = \frac{\sum(xy) - a_0\sum(x)}{\sum(x^2)}$$

QUADRATIC LEAST SQUARES:

$$y = a_0 + a_1 x + a_2 x^2$$

$$\begin{aligned}
a_0 N \quad &+ a_1\sum(x) + a_2\sum(x^2) = \sum(y) \\
a_0\sum(x) &+ a_1\sum(x^2) + a_2\sum(x^3) = \sum(xy) \\
a_0\sum(x^2) &+ a_1\sum(x^3) + a_2\sum(x^4) = \sum(x^2 y)
\end{aligned}$$

Matrix Representation:

$$\mathbf{AX} = \mathbf{B}, \quad \mathbf{X} = \mathbf{A}^{-1}\mathbf{B}$$

$$\mathbf{A} = \begin{bmatrix} N & \sum(x) & \sum(x^2) \\ \sum(x) & \sum(x^2) & \sum(x^3) \\ \sum(x^2) & \sum(x^3) & \sum(x^4) \end{bmatrix}, \quad \mathbf{X} = \begin{bmatrix} a_0 \\ a_1 \\ a_2 \end{bmatrix}, \quad \mathbf{B} = \begin{bmatrix} \sum(y) \\ \sum(xy) \\ \sum(x^2 y) \end{bmatrix}$$

Analytically:

$$A_1 = N\sum(x^2) - \sum(x)\sum(x)$$
$$A_2 = N\sum(x^3) - \sum(x)\sum(x^2)$$
$$A_3 = N\sum(x^4) - \sum(x^2)\sum(x^2)$$

$$B_1 = N\sum(xY) - \sum(x)\sum(y)$$
$$B_2 = N\sum(x^2 Y) - \sum(x^2)\sum(y)$$

$$a_2 = \frac{A_1 B_2 - A_2 B_1}{A_1 A_3 - A_2 A_2}$$

$$a_1 = \frac{B_1 - a_2 A_2}{A_1}$$

$$a_0 = \frac{\sum(xy) - a_1\sum(x^2) - a_2\sum(x^3)}{\sum(x)}$$

CUBIC LEAST SQUARES:

$$y = a_0 + a_1 x + a_2 x^2 + a_3 x^3$$

$$\begin{aligned}
&a_0 N + a_1 \sum(x) + a_2 \sum(x^2) + a_3 \sum(x^3) = \sum(y) \\
&a_0 \sum(x) + a_1 \sum(x^2) + a_2 \sum(x^3) + a_3 \sum(x^4) = \sum(xy) \\
&a_0 \sum(x^2) + a_1 \sum(x^3) + a_2 \sum(x^4) + a_3 \sum(x^5) = \sum(x^2 y) \\
&a_0 \sum(x^3) + a_1 \sum(x^4) + a_2 \sum(x^5) + a_3 \sum(x^6) = \sum(x^3 y)
\end{aligned}$$

Matrix Representation:

$$\mathbf{AX} = \mathbf{B}, \qquad \mathbf{X} = \mathbf{A}^{-1}\mathbf{B}$$

$$\mathbf{A} = \begin{bmatrix} N & \sum(x) & \sum(x^2) & \sum(x^3) \\ \sum(x) & \sum(x^2) & \sum(x^3) & \sum(x^4) \\ \sum(x^2) & \sum(x^3) & \sum(x^4) & \sum(x^5) \\ \sum(x^3) & \sum(x^4) & \sum(x^5) & \sum(x^6) \end{bmatrix}, \quad \mathbf{X} = \begin{bmatrix} a_0 \\ a_1 \\ a_2 \\ a_3 \end{bmatrix}, \quad \mathbf{B} = \begin{bmatrix} \sum(y) \\ \sum(xy) \\ \sum(x^2 y) \\ \sum(x^3 y) \end{bmatrix}$$

Analytically:

Analytical solution of the above equations is impractical when matrix methods are available.

LEAST SQUARES RELATIONSHIPS CHARACTERISTICS

The table below generalizes the characteristics of the least squares relationships developed above.

Order of the least squares polynomial curve fit, n	Minimum number of data points needed to calculate, N	Size of the $\mathbf{A}$ matrix and the $\mathbf{A}^{-1}$ matrix	Size of the $\mathbf{B}$ vector	Size of the $\mathbf{X}$ vector	Highest power of x's in the x sums in the $\mathbf{A}$ matrix needed to calculate	Highest power of x's in the xy product sums in the $\mathbf{B}$ vector needed to calculate
1	3	2x2	2x1	2x1	2	1
2	4	3x3	3x1	3x1	4	2
3	5	4x4	4x1	4x1	6	3

n	$n+2$	$(n+1)\mathrm{x}(n+1)$	$(n+1)\mathrm{x}1$	$(n+1)\mathrm{x}1$	$2n$	n

Table A3.1

For example, the table tells us that, if we want to fit a set of data points with a third order (cubic) polynomial ($n=3$) by the method of least squares, we will need to have at least five (5) data points (N), the size of the $\mathbf{A}$ matrix needed to make the calculations will be 4x4, the inverted $\mathbf{A}$ matrix ($\mathbf{A}^{-1}$) will also be 4x4 in size, the right-hand-side vector ($\mathbf{B}$) and the solution vector ($\mathbf{X}$) will both be 4x1 in size, the $\mathbf{A}$ matrix will contain the sums of the x's up to the sixth power (x^6), and the right-hand-side vector ($\mathbf{B}$) will contain sums of the xy products up to x's to the third power (x^3y).

APPENDIX 4: LINEAR (STRAIGHT-LINE) FIT OF TWO POINTS

The following is an outline showing how to fit a linear or straight-line, first-order polynomial, to two given data points.

Given the data points $\left(x_1, y_1\right), \left(x_2, y_2\right)$;

For $y = a_0 + a_1 x$:

$$a_1 = \frac{y_2 - y_1}{x_2 - x_1}$$

$$a_0 = y_2 - a_1 \left(x_2 \right) = y_1 - a_1 \left(x_1 \right)$$

Example:

Given the data points $\left(2,2\right), \left(3,4\right)$. In this example, $x_1 = 2, x_2 = 3, y_1 = 2$, and $y_2 = 4$. Then, to fit an equation of the form $y = a_0 + a_1 x$ to the two points:

$$a_1 = \frac{y_2 - y_1}{x_2 - x_1} = \frac{4 - 2}{3 - 2} = \frac{2}{1} = 2$$

$$a_0 = y_2 - a_1 \left(x_2 \right) = 4 - 2(3) = -2$$

$$a_0 = y_1 - a_1 \left(x_1 \right) = 2 - 2(2) = -2$$

So that the straight-line equation that fits the two points is $y = -2 + 2x$.

Using Microsoft® Excel to make and confirm the calculations, where y_{calc} are the values of y calculated from the linear equation:

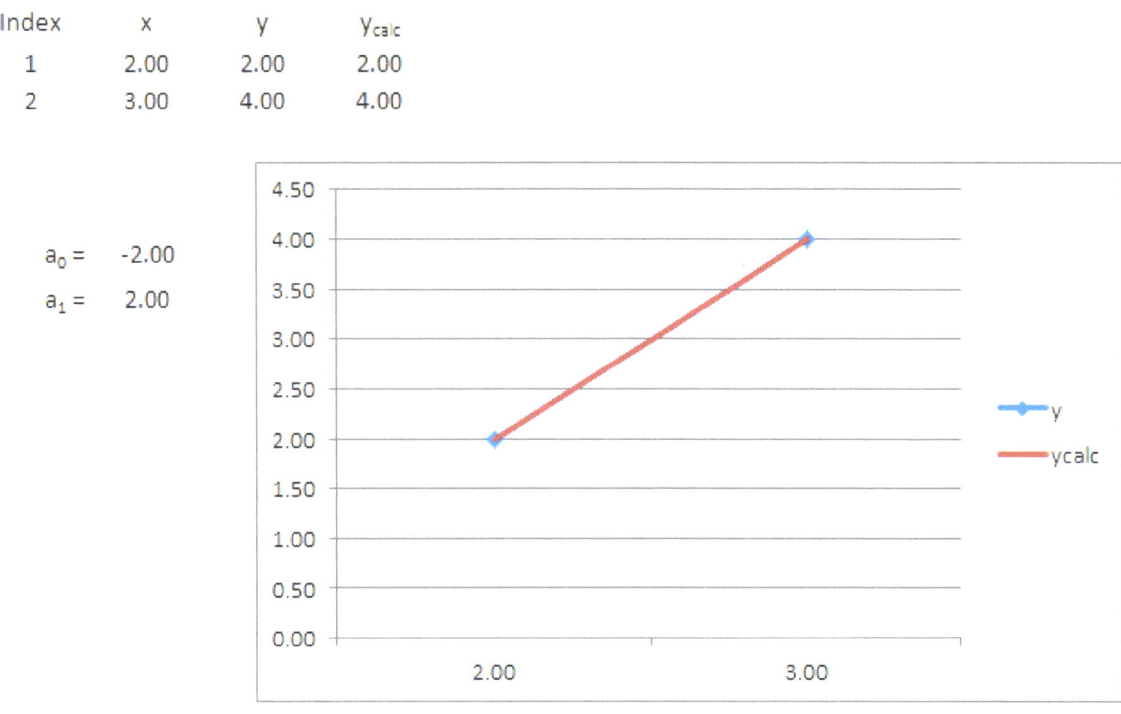

Index	x	y	y_{calc}
1	2.00	2.00	2.00
2	3.00	4.00	4.00

$a_0 =$ -2.00

$a_1 =$ 2.00

Figure A4.1

APPENDIX 5: QUADRATIC FIT OF THREE POINTS

The following is an outline showing how to fit a quadratic, second-order polynomial, to three given data points.

Given the data points $(x_1, y_1), (x_2, y_2), (x_3, y_3)$;

For $y = a_0 + a_1 x + a_2 x^2$:

$$a_2 = \frac{1}{x_3 - x_2}\left[\frac{y_3 - y_1}{x_3 - x_1} - \frac{y_2 - y_1}{x_2 - x_1}\right]$$

$$a_1 = \frac{y_3 - y_1}{x_3 - x_1} - a_2\left(x_3 + x_1\right)$$

$$a_0 = y_2 - a_2\left(x_2\right)^2 - a_1\left(x_2\right)$$

Example:

Given the data points $(2,2), (3,4), (4,5)$. In this example, $x_1 = 2$, $x_2 = 3$, $x_3 = 4$, $y_1 = 2$, $y_2 = 4$, and $y_3 = 5$. Then, to fit an equation of the form $y = a_0 + a_1 x + a_2 x^2$ to the three points:

$$a_2 = \frac{1}{x_3 - x_2}\left[\frac{y_3 - y_1}{x_3 - x_1} - \frac{y_2 - y_1}{x_2 - x_1}\right]$$

$$= \frac{1}{4 - 3}\left[\frac{5 - 2}{4 - 2} - \frac{4 - 2}{3 - 2}\right] = \frac{1}{1}\left[\frac{3}{2} - \frac{2}{1}\right] = -\frac{1}{2} = -0.5$$

$$a_1 = \frac{y_3 - y_1}{x_3 - x_1} - a_2\left(x_3 + x_1\right) = \frac{5 - 2}{4 - 2} - \left(-\frac{1}{2}\right)\left(4 + 2\right)$$

$$= \frac{3}{2} + \left(\frac{1}{2}\right)\left(6\right) = \frac{3}{2} + 3 = \frac{9}{2} = 4.5$$

$$a_0 = y_2 - a_2\left(x_2\right)^2 - a_1\left(x_2\right) = 4 - \left(-\frac{1}{2}\right)\left(3\right)^2 - \left(\frac{9}{2}\right)\left(3\right)$$

$$= 4 + \frac{9}{2} - \frac{27}{2} = -\frac{10}{2} = -5$$

So that the quadratic equation that fits the three points is

$y = -5.0 + 4.5\,x - 0.5\,x^2$.

Using Microsoft® Excel to make and confirm the calculations, where y_{calc} are the values of y calculated from the quadratic equation:

Index	x	y	y_{calc}
1	2.00	2.00	2.00
2	3.00	4.00	4.00
3	4.00	5.00	5.00

a_0 =	-5.00
a_1 =	4.50
a_2 =	-0.50

Figure A5.1

APPENDIX 6: BIBLIOGRAPHY, SOFTWARE, URL'S

1. MathType, Version 6.8, Copyright © 1990-2012 Design Science, Inc.

2. Microsoft® Excel 2010, Version 14.0.6129.5000 (32-bit), Part of Microsoft® Office Professional 2010, Copyright © 2010 Microsoft® Corporation

3. Microsoft® Word 2010, Version 14.0.6129.5000 (32-bit), Part of Microsoft® Office Professional 2010, Copyright © 2010 Microsoft® Corporation

4. Mylander, Charles, “Secrets for Growing Churches”, Harper& Row, First Edition (1979), pages 107-108, ISBN 0-06-066055-4

5. Snagit, The Windows Screen Capture Utiltity, Version 11.2.0, Copyright © 1996-2013 TechSmith Corp.

6. TheFreeDictionary, http://www.thefreedictionary.com/skinny, April 9, 2013

7. Wikipedia, The Free Encyclopedia, http://en.wikipedia.org/wiki/Least_squares, April 9, 2013

GLOSSARY

Array – a systematic arrangement of numbers or objects arranged in rows and columns.

COUNT function – a mathematical procedure in Microsoft® Excel that counts the number of items in a range of Excel cells.

Cubic Polynomial - a mathematical expression of the form $y = a_0 + a_1 x + a_2 x^2 + a_3 x^3$, where x represents the independent variable, normally plotted along the horizontal axis of a rectangular coordinate system, y represents the dependent variable, normally plotted along the vertical axis of a rectangular coordinate system, the a_n's are constants, and the exponents of x are positive integers.

Curve fit – an analytical mathematical expression that represents a set of data points.

Dependent variable – the output from a mathematical expression, frequently represented by the letter y and normally plotted along the vertical axis of a rectangular coordinate system

Differential Calculus – a part of calculus that deals with the rate at which quantities change.

Excel – a spreadsheet application developed by the Microsoft® Corporation.

Extrapolate/extrapolation - the process of finding a value beyond or outside of the domain of a given set of data.

Independent variable – the input to a mathematical expression, frequently represented by the letter x and normally plotted along the horizontal axis of a rectangular coordinate system.

Interpolate/interpolation - the process of finding a value within the domain of a given set of data.

Least squares – a mathematical criterion that minimizes the sum of the squares of the vertical distances of a set of data points from a proposed analytical representation of those data.

Linear Polynomial - a mathematical expression of the form $y = a_0 + a_1 x$, where x represents the independent variable, normally plotted along the horizontal axis of a rectangular coordinate system, y represents the dependent variable, normally plotted along the vertical axis of a rectangular coordinate system, the a_n's are constants, and the exponents of x are positive integers.

Matrix – a rectangular array of numbers arranged in rows and columns. The items in a matrix are elements or entries.

Microsoft® – registered trademark of Microsoft Corporation.

MINVERSE function – a mathematical procedure in Microsoft® Excel that calculates the inverse of a given matrix.

MMULT function – a mathematical procedure in Microsoft® Excel that calculates the product of two matrices or of a matrix and a vector.

Polynomial – a mathematical expression of the form $y = a_0 + a_1 x + a_2 x^2 + \ldots + a_n x^n$, where x represents the independent variable, normally plotted along the horizontal axis of a rectangular coordinate system, y represents the dependent variable, normally plotted along the vertical axis of a rectangular coordinate system, the a_n's are constants, and the exponents of x are positive integers.

Quadratic Polynomial - a mathematical expression of the form $y = a_0 + a_1 x + a_2 x^2$, where x represents the independent variable, normally plotted along the horizontal axis of a rectangular coordinate system, y represents the dependent variable, normally plotted along the vertical axis of a rectangular coordinate system, the a_n's are constants, and the exponents of x are positive integers.

Rectangular coordinates – a coordinate system that uniquely specifies each point in a plane by a pair of numerical values, each measured in the same unit of length from the intersection of two fixed perpendicular directed lines. Also called Cartesian coordinates.

Regression – a statistical analysis technique for estimating the relationship among variables.

SOLVER add-in – a 'what-if' analysis tool in Microsoft® Excel that performs optimization calculations.

Spreadsheet – an interactive computer application for calculating, analyzing, and reporting data arranged in tabular form. Data are arranged vertically in columns and horizontally in rows. The intersection of a column and a row is called a cell.

Statistics/statistical – the mathematics of the collection, organization, and interpretation of numerical data.

SUM function – a mathematical procedure in Microsoft® Excel that calculates the sum of the values contained in a range of Excel cells.

Vector – a one-column array of numbers.

INDEX

M

N

O

P

Q

R

S

T

U

V

W

Z

www.ingramcontent.com/pod-product-compliance
Lightning Source LLC
Chambersburg PA
CBHW051018180526
45172CB00002B/389
9781483919300